UNM - GALLUP

ZOLLINGER LIBRARY

200 COLLEGE ROAD

GALLUP, NEW MEXICO   87301

1.   Books may be kept two weeks and may be renewed once for the same period.

2.   A fine is charged for each day a book is not returned according to the above rule.   No book will be issued to any person incurring such a fine until it has been paid.

3.   All injuries to books beyond reasonable wear and all losses shall be made good to the satisfaction of the Librarian.

4.   Each borrower is held responsible for all books charged on his card and for all fines accruing on the same.

Animal Osmoregulation

J. C. RANKIN, Ph.D.

Lecturer in Zoology
University College of North Wales

J. DAVENPORT, Ph.D.

Principal Scientific Officer
N.E.R.C. Unit, Marine Science Laboratories
University College of North Wales

A HALSTED PRESS BOOK

John Wiley and Sons

New York and Toronto

Blackie & Son Limited
Bishopbriggs
Glasgow G64 2NZ

Furnival House
14–18 High Holborn
London WC1V 6BX

Published in the U.S.A. and Canada
by Halsted Press, a Division of
John Wiley & Sons, Inc., New York

Library of Congress Cataloging in Publication Data
Rankin, J. C.
  Animal osmoregulation.
  (Tertiary level biology series)
  "A Halsted Press book."
  Bibliography: p.
  Includes index.
  1. Osmoregulation.
  I. Davenport, J. II. Title. III. Series.
  QP90.6.R36 1981   591.1'88   81–7491
  ISBN 0–470–27207–4   AACR2

Filmset by Advanced Filmsetters (Glasgow) Ltd.

Printed in Great Britain by
Thomson Litho Ltd, East Kilbride, Scotland

Preface

LIFE AS WE KNOW IT ON EARTH BEGAN IN A DILUTE AQUEOUS SALT solution, the sea. Animals have been able to colonise other environments only by evolving the ability to maintain dilute aqueous salt solutions within their bodies. Most of the body of an animal consists of these solutions, contained by cell membranes or (in the case of the extracellular fluids of multi-cellular animals) epithelial cell layers. Osmoregulation is the regulation of their solute concentrations, or, to express it in a different way, their water activities. Animals possess osmoregulatory capability to varying degrees. Many marine animals which are incapable of regulating the osmotic concentrations of their extracellular fluids can survive only in sea water. At the other extreme, some animals overcome the great problems of water or salt shortage or excess imposed by habitats as diverse as waterless deserts, almost salt-free fresh water, or saturated salt solutions.

The obvious consequences of osmoregulation are that they determine the sort of environment an animal can live in, but the processes involved are not usually directly apparent, and measurement techniques are therefore of crucial importance; the study of osmoregulation combines aspects of both biology and chemistry. For this reason we feel it essential, even in a short book, to pay particular attention to experimental details. This inevitably means that some examples, selected to illustrate particular aspects of osmoregulation, are discussed at some length, whilst other sections of the book are more general. In view of the widespread occurrence of parallel evolution, with the adoption of similar solutions to common problems, we have not attempted a comprehensive coverage of osmoregulation in all animal groups.

The book is arranged in relation to the problems of animals which have left the sea for brackish water, fresh water or land, those in the last category having to overcome the severe problems of regulating body water and salt content when no longer permanently immersed in dilute salt solutions. The examples chosen for more detailed consideration may reflect our own personal interests, but the principles they illustrate are of universal application in the study of osmoregulation.

Many excellent textbooks on osmoregulation have appeared, from Krogh (1939) to Potts and Parry (1964), but the vast amount of data which has accumulated in more recent years has not been satisfactorily reviewed and presented in a form intelligible to those seeking an introduction to the subject. The tendency has been towards the production of excellent (and expensive!) multi-author treatises, in which specialists have reviewed various aspects of the subject in great detail. In contrast, there are many comparative physiology textbooks, notably that of Schmidt-Nielsen (1979), which have interesting and informative sections on osmoregulation but, because of limitations of space, can describe only some of the fascinating aspects of the subject, or else generalise and avoid recent advances and controversial issues.

We feel that there is a need for a book to fill the gap between these two approaches, and which will give sufficient detail, by the use of representative examples, to indicate the flavour of current research in osmoregulation, whilst presenting a balanced view of the subject and pointing the interested reader in the direction of more specialised works. Many topics are interesting and stimulating because their interpretation is controversial, often because existing experimental techniques are inadequate. We feel that it is an essential part of a scientific education to appreciate that scientific knowledge evolves continually, and that it is just as important to understand how facts are obtained and used to formulate theories, as it is to memorise the facts and theories themselves. We hope that we will assist a few of our readers to a deeper understanding of the subject, which may lead them to question currently fashionable concepts and possibly go on to replace them with more original ideas based on their own researches.

We have deliberately included only a few references (chosen to make particular points) in the text, because, for readers wishing to pursue any one topic, the plethora of recent review articles (listed under Further Reading) provides excellent detailed information. We hope that our many colleagues who have not been given proper credit for their important contributions will forgive us. We are most grateful to a number of people, especially Valmai Griffiths for help with typing and drawing figures, Inger Wahlqvist and Dr D. J. Patterson for drawing figures, Julia Davenport for compiling the index, and to Professor P. J. Bentley and Dr G. Wyn Jones for reading parts of the manuscript and making useful suggestions.

J.C.R.
J.D.

Contents

CHAPTER ONE

BASIC PRINCIPLES

Introduction

Life on earth is thought to have begun in the sea. Organic compounds, possibly produced by lightning discharges in the primitive atmosphere (which is now thought to have consisted mainly of carbon dioxide and water vapour), or even, according to some theories, of extraterrestrial origin, dissolved in the sea and reacted with each other. Self-replicating compounds must eventually have formed and the process of evolution started. All the chemical reactions of life would have taken place in the dilute salt solution of the early seas until cell membranes evolved. Thereafter, although the composition of the internal fluids of the primitive organisms could be modified, they remained dilute salt solutions. The dilute sodium chloride solution flowing through our veins today is a reminder of our ancient marine origin.

In the course of evolution many groups of animals successfully made the transition from sea via brackish-water estuaries to fresh water, and some emerged on to land. The way in which the composition of the body fluids is regulated in animals inhabiting these different environments is the subject of this book. There are many fascinating examples of animals which encounter and overcome extreme osmoregulatory problems—from salmon migrating between the sea and fresh water, to camels crossing waterless deserts. Before we consider some of these, however, we must define what we mean by osmoregulation, and consider the basic principles underlying the movement of water and ions across cell membranes and epithelia. *Osmoregulation may be defined as regulation of the osmotic concentration of the body fluids.* We will now consider this definition in more detail before discussing the ways in which animals osmoregulate.

The body fluids

The bulk of the body of any animal consists of aqueous solutions of various solutes. Parts of the body may be solid and crystalline (bone, shells,

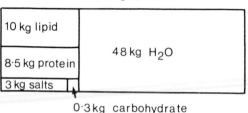

Figure 1.1   What the human body is made of.

exoskeletons etc.), and parts are composed of lipid from which water is excluded, but water is the main constituent of all organisms. The human body, for example, contains 45–75 % water. (The reason for such great variation between individuals is simple—fat deposits, which contain no water, may be present to a greater or lesser extent; the water content of any one individual, however, remains constant.) The proportion of water relative to the other principal components of the human body is shown diagrammatically in figure 1.1. The way in which the total body water can be measured, together with the water content of the various sub-compartments, will be described in chapter 11. Obviously evaporation to dryness is not a suitable technique!

The higher figure quoted above is typical of most normal animals. The body water of unicellular animals must all be intracellular, but in higher forms a significant amount of extracellular water also exists. This may include fluid in circulatory systems and in various body cavities, secretions of glands etc., but the bulk of it is interstitial fluid bathing all the cells of the body.

As the great nineteenth-century French physiologist Claude Bernard pointed out, the extracellular fluid constitutes an "internal environment". Maintenance of a constant internal environment enables the cells to function more efficiently in that they are not affected by the nature of the external environment. Animals can be classified as "osmoregulators" or "osmoconformers" depending on whether they can or cannot maintain the osmotic concentration of their extracellular fluid constant in the face of changes in external osmotic concentrations. Some single-celled animals can osmoregulate, but apart from lacking a compartment which can shield the cells from the vicissitudes of the external environment, they also suffer from the great disadvantage of a high surface area to volume ratio which favours exchange with the external medium.

Water

It is easy to demonstrate that living organisms contain a lot of water, but proving that this water is in the same physical state as pure liquid water, and that the solute molecules present in it are in true solution, is much more difficult. Arguments about the state of water in cells have continued for many years and the situation is complicated by the fact that we still do not know the exact physical state of pure water. This is because water is a most abnormal liquid, because of the tendency of water molecules to form hydrogen bonds with each other. In the solid state this leads to the formation of a regular crystalline lattice (in fact eight different types of lattice occur under differing conditions of temperature and pressure; these have been named ice I to ice VIII). In the liquid state water molecules are more tightly packed than in ice and they retain their propensity to form hydrogen bonds. It has been suggested that water contains clusters of ice I and ice III crystals or alternatively that it consists of two interpenetrating partially occupied ice VIII lattices. More recent theories include the "flickering cluster" model, in which small hydrogen-bonded groups of molecules continually form and re-form, and the gel model, which postulates an infinitely and randomly changing pattern of hydrogen bonding throughout the liquid. The problem which has prevented elucidation of the structure of liquid water to date is the extremely short lifetime ($< 10^{-11}$ s) of any of the proposed structures compared to the observation times of any of the techniques in use (e.g. $> 10^{-10}$ s for X-ray diffraction).

In the presence of ions, water molecules are orientated into shells of water of hydration and the greater the charge on the surface of the ion the more pronounced the ordering of the surrounding water molecules. The cations of the alkali metals lithium, sodium, potassium, rubidium and caesium are all univalent, i.e. have an excess of one proton over the number of electrons, but because the size of the electron cloud separating the positive nucleus from the water increases with increasing atomic weight, we find that lithium has the most water of hydration and caesium the least. In fact the hydrated lithium ion is larger than the hydrated sodium ion, and so on until the hydrated caesium ion is the smallest of the series.

Water molecules tend to form ordered arrays adjacent to solid surfaces and also, presumably, at cell membranes. Differential scanning calorimetry experiments on membrane phospholipids in the presence of small amounts of water fail to produce evidence of melting of ice as the temperature is raised through 0°C, implying that the water has the same structure above and below this temperature. These layers are probably only a few molecules thick, but cells contain many membranes and many ions and charged protein molecules.

It has been suggested that the majority of the water molecules in cells are not in the liquid state but form a gel in which solute molecules are trapped The latest work, using techniques such as nuclear magnetic resonance to determine how free water molecules are to rotate, suggests that the bulk of the intracellular water is in the same physical state as pure liquid water (whatever that may be) and that only a few per cent of the water molecules are irrotationally bound. However, molecules diffuse within cells at only about half the rate they do in pure water, probably because the presence of so many membranes and macromolecules makes diffusion paths longer than they appear.

In spite of some evidence to the contrary we will continue, with some exceptions, to use the gross oversimplification employed by generations of physiologists and regard cells as bags full of fluid containing solutes in simple solution. The composition of the cell is thus determined by what passes in or out across the cell membrane, although of course metabolic activity can alter constituents within the cell. This simple approach has proved adequate to explain cellular osmoregulation—the regulation of intracellular osmotic concentrations.

Osmotic concentrations

Revision of some simple chemistry may be useful at this stage. When a solution is separated from a pure solvent by a membrane which is impermeable to the solute (a *semipermeable membrane*) solvent passes into the solution. This movement can be prevented by the application of the correct hydrostatic pressure to the solution—the *osmotic pressure* of the solution. It was demonstrated by van 't Hoff that osmotic pressure was equal to nRT/V, where n is the number of gram moles of solute, R the gas constant, T the temperature in kelvins and V the volume in litres. He pointed out that this expression is analogous to the ideal gas law. In other words, 1 mole of solute dissolved in 22.4 l of solution at 273°K would appear to be exerting a negative pressure of 1 atmosphere, pulling solvent into the solution. An alternative way of looking at osmosis is to consider that the presence of solute molecules lowers the *chemical potential* of solvent in solution. An example will help to illustrate what is meant by this. In figure 1.2 the pure water on the left of the semi-permeable membrane has a concentration of $1000 \, \text{g} \, \text{l}^{-1} = 55.5 \, \text{mol} \, \text{l}^{-1}$ (1000/18) and we will assume that its activity coefficient $= 1$. The presence of solute molecules on the right-hand side reduces the water activity, so there will be a net movement

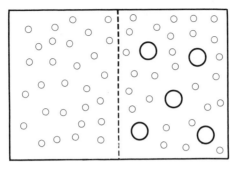

Figure 1.2 Osmosis across a semi-permeable membrane (broken line). Small circles represent water molecules, large circles solute molecules.

of water molecules from left to right, down their activity gradient.* This flow can be prevented by raising the chemical potential of the water molecules in the solution. This could be achieved by raising their temperature above that of pure water, although in practice conduction of heat across the membrane would make it difficult to maintain a temperature gradient. The same effect could be produced by the application of pressure to the solution, provided that the membrane was rigid. In this case the gradient of chemical potential (u_w) would be related to the pressure gradient (Δp) and the gradient of water activity. Since it is difficult to determine the water activity, the mole fraction of water (n_w—the number of water molecules divided by the number of water + solute molecules per unit volume) can be used instead, in the following equation:

$$u_w = \overline{V}_w \Delta p + RT\Delta(\ln n_w)$$

$\overline{V}$ is the partial molar volume of water ($cm^3\ mol^{-1}$), R is the gas constant and T the absolute temperature.

For most practical purposes it does not matter which view of the nature of osmosis is taken, but it is preferable to talk about osmotic concentrations rather than osmotic pressures which exist only in the laboratory, in theory or in plants (almost never in animals, where as soon as an osmotic gradient is developed water moves down it, since animal cells can tolerate only small hydrostatic pressure gradients.) In almost all instances

---

* As can be seen from Table 1.1 an electrolyte reduces the water concentration but does not exactly displace its own volume of water, and the formation of ordered shells of water of hydration also affects the activity coefficient of the water, so the equation which follows is accurate only at infinite dilution.

which have been studied in detail in the animal kingdom, movements of water have been found to be passive flows of water down gradients of water chemical potential. *Active transport** of water used to be invoked to explain a number of fluid movements, for example the reabsorption of water needed to produce concentrated urine in the mammalian kidney. Here, it was later found that a complicated anatomical arrangement, involving the loops of Henle, the collecting ducts and the vasa recta blood vessels had evolved to create local osmotic gradients along which water moved. In other cases, where fluid movement across epithelia appeared to be from regions of low to high water potentials, the creation of local osmotic and hydrostatic pressure gradients on a microscopic scale served to ensure passive movement of water, contrary to the overall osmotic gradients. In a few instances, notably rectal absorption of water vapour in insects, the maximum osmotic concentrations which could be achieved, even by saturated salt solutions, would be insufficient to cause osmotic absorption of water in some of the situations studied, and active transport of water remains a possibility.

The way in which animals regulate water movement across cell membranes and epithelia will be discussed in detail later, but we must bear in mind from the start that osmoregulation is essentially regulation of solute concentrations, and that water movement is almost always passive. Animals rarely maintain osmotic pressure differences balanced by hydrostatic pressures as plants do (one exceptional case is discussed in chapter 10). When differences in osmotic concentrations across membranes are found either the membrane is impermeable to water (e.g. the mammalian collecting duct membrane in the absence of antidiuretic hormone) or water is passing down its chemical gradient (e.g. the same membrane in the presence of antidiuretic hormone).

Units

The osmotic concentration of a 1 molar solution of an ideal solute is said to be 1 osmolar. It is often preferable to use molal and osmolal units. A 1 molar solution contains 1 mole of solute per 1 of solution; a 1 molal solution contains 1 mole of solute dissolved in 1 kg solvent. For dilute solutions the difference is slight but it becomes greater at increased concentrations, as shown for sodium chloride in Table 1.1. The concentration of the solute in question need not be high for a difference to exist

---

* Active transport is defined as movement of a substance against its gradient of chemical, or in the case of charged molecules, electrochemical potential. Energy must be supplied—in biological systems usually from hydrolysis of ATP.

Table 1.1 Sodium chloride solutions

$\%_0$	$g\,l^{-1}$	$mol\,l^-$	$mol\,kg^{-1}$	$Osm\,kg^{-1}$	$\Delta°C$	$[H_2O]\,mol\,l^{-1}$
1	1.0	0.017	0.017	0.033	0.062	55.4
5	5.0	0.086	0.086	0.161	0.299	55.4
10	10.1	0.172	0.173	0.319	0.593	55.3
20	20.2	0.346	0.349	0.638	1.186	55.1
30	30.6	0.523	0.529	0.962	1.790	54.9
40	41.1	0.703	0.713	1.295	2.409	54.8
50	51.7	0.885	0.901	1.638	3.046	54.6
100	107.1	1.832	1.901	3.529	6.564	53.5
150	166.3	2.845	3.020	5.854	10.888	52.3

$\%_0$ = parts per thousand by weight
$\Delta°C$ = depression of freezing point in °C
$[H_2O]$ = concentration of water in solution

Source: Weast, R. C. (1978–79) CRC Handbook of Chemistry and Physics, CRC Press Inc., Palm Beach.

between molar and molal units—it is the total solute concentration which matters. For example, in mammalian plasma containing 8 % solids (mainly plasma proteins) the sodium concentration may be 140 mmolar $(mmol\,l^{-1})$ or 152 mmolal $(mmol\,kg^{-1})$.

The osmolal concentration is only equal to the molal concentration for an ideal solute and an ideal semi-permeable membrane. These conditions are lacking in most biological situations. Since osmotic concentration is a colligative property (depending on the number of solute particles per unit volume) high molecular weight substances such as proteins contribute little to the osmotic concentrations of most body fluids even though they may make up the greater part of the mass of solute dissolved. Small molecules, particularly salts, make up most of the osmotic concentration. Salts of strong acids dissociate completely in solution, so a 1 molal sodium chloride solution might be expected to be 2 osmolal. The measured osmolality (Table 1.1) is in fact less than would be expected from the number of particles in solution. This is not because of incomplete dissociation (as is found with salts in weak acids), but is due to interactions between positive and negative ions in solution. On average, a given cation will have one more negative than positive ion in its immediate vicinity at any point in time so will not be as free to move as it would be in the absence of the attraction of the surrounding anions. The salt therefore behaves as if it were not completely dissociated in solution.

The second condition, an ideal semi-permeable membrane, is also never satisfied, because all biological membranes are permeable to salts to some extent. A correction factor, the *Staverman* or *reflection coefficient*, σ, has to be introduced:

measured osmotic pressure = σ (theoretical osmotic pressure)

The coefficient σ can vary from 1 (ideal semi-permeable membrane) to 0 (membrane as permeable to solute as to solvent molecules). In practice it is often assumed to be equal to 1, as biological membranes are usually very much less permeable to salts than to water, and in any case it is difficult to measure.

If two solutions have the same osmotic concentrations they are said to be iso-osmotic. If one solution has a greater osmotic concentration than another the former is said to be hyperosmotic and the latter hypo-osmotic. The terms isotonic, hypertonic and hypotonic are often used and are frequently confused with iso-osmotic, hyperosmotic and hypo-osmotic. A solution is isotonic to a cell or piece of tissue if the cell or tissue does not shrink or swell when placed in that solution. The solution may not necessarily have the same osmotic pressure as the fluid within the cell or cells, as will be explained below. Unfortunately, since "iso-osmotic" is a rather clumsy word, "isotonic" is often used rather indiscriminately.

A common way of expressing concentrations is parts of solute per thousand parts of the solution by weight (‰). In the case of sea water the

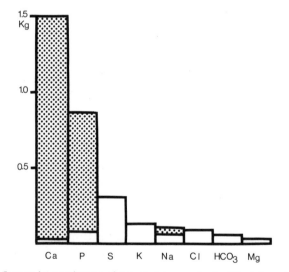

Figure 1.3 Inorganic constituents of a typical human body. Stippled parts of columns represent amounts present in bone.

total solute concentration in ‰ is referred to as the *salinity*. We will use this term, but in general will use molal and osmolal concentrations where possible. The salinity of sea water is usually around 34.5‰ which is exactly $1.000 \text{ Osm kg}^{-1}$. As most biological fluids are at or below this osmolality it is convenient to use milliosmoles (mOsm) and millimoles (mmol) throughout. In some instances, when considering the balance between anions and cations, it is clearer to use milliequivalents rather than millimoles.

Osmotic pressures can be measured directly, but it is often more convenient and accurate to estimate osmotic concentrations by measurement of some other colligative property, for example depression of solute vapour pressure or, more commonly, depression of the freezing (or melting) point. In fact depression of freezing point of water, in °C, is often quoted as a measurement of osmotic concentration. Freezing point depressions for sodium chloride solutions are given in Table 1.1.

Regulation of osmotic concentrations

The body fluids of all living organisms are dilute salt solutions reflecting the origin of life in the sea. They also contain organic solutes, of course, although Ringer, in 1883, demonstrated that frog hearts would continue to beat normally for long periods in a solution containing only the chlorides of sodium, potassium and calcium in the correct concentrations, and physiological salt solutions have been called "Ringer's solutions" or just "Ringer" ever since.

The inorganic salts contain the elements shown in figure 1.3, where the amounts depicted are those present in the average human body. Bicarbonate is also included, as it is convenient to consider it as if it was an inorganic salt although it is subject to continuous turnover as part of the respiratory process.

The cross-hatched parts of the columns represent the amount present in the bone—most of the body's calcium and phosphorus and also a significant amount of its sodium. Much of the remainder of the calcium and phosphorus, most of the sulphur and a good deal of the magnesium are in combination with organic compounds. The salts which contribute to the osmotic concentration of the body fluids are mainly the chlorides and the phosphates and bicarbonates (in air-breathing animals) of sodium and potassium. These are not distributed evenly between the intracellular and extracellular fluids. Figure 1.4 shows the ionic concentrations in mammalian skeletal muscle cells and the interstitial fluid surrounding

INTRACELLULAR   EXTRACELLULAR

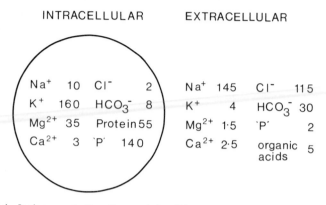

Figure 1.4   Ionic concentrations (in mequiv kg^{-1}) in a mammalian skeletal muscle cell water and in the interstitial fluid surrounding it.

them. Concentrations are given in mequiv kg^{-1} to show how the anions and cations balance each other, but since protein and phosphate molecules are polyvalent their contributions to intracellular osmotic concentrations are much lower than their concentrations suggest.

Minor components have been omitted, so positive and negative charges do not balance exactly, but ions making up most of the osmotic concentrations of mammalian body fluids, of about 300 mOsm kg^{-1}, have been included.

In many marine animals the osmotic pressure of the extracellular fluids is identical to that of the environment at all times, i.e. they are *osmo-conformers*, not *osmoregulators*. They still maintain similar intracellular ionic concentrations to those shown in figure 1.4, adjusting the total osmotic concentration by synthesis or degradation of organic compounds, as will be discussed in chapter 3.

The intracellular potassium concentration shows a remarkable constancy in all animals, being between 100 and 200 mmolal in all but some of the less advanced groups of freshwater animals which have evolved tolerance of very dilute body fluids. Figure 1.5 illustrates the fact that all animals with an extracellular fluid sodium concentration of more than 140 mmol l^{-1} can maintain intracellular potassium concentrations of around 160 mmol l^{-1}.

How are the differences in salt concentrations between intracellular and extracellular fluid maintained? Firstly, some compounds are present at high concentrations within the cell because they cannot readily penetrate the cell membrane, e.g. proteins, organic phosphates, amino acids, protein-

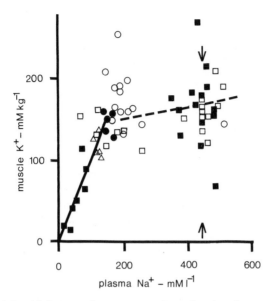

Figure 1.5 Relationship between plasma, serum or haemolymph sodium concentrations (in mmol l^{-1}) and muscle cell potassium concentrations (in mmol kg^{-1} cell water) in arthropods ($\square$), other invertebrates ($\blacksquare$), fish, including *Myxine* ($\bigcirc$), Amphibia ($\triangle$), reptiles, birds and mammals ($\bullet$). Redrawn from Burton (1968).

bound magnesium. But sodium, potassium, chloride and bicarbonate ions can pass across cell membranes. The distribution which results from the presence of permeant and non-permeant ions in such a situation is known as a *Donnan equilibrium*. If we consider a two-compartment system divided by a membrane permeable to sodium and chloride ions but not to proteins, and which contains the number of particles shown in figure 1.6 (although in reality each protein molecule would have a number of charged groups), how many sodium and chloride ions would be required in compartment B to ensure equilibrium?

Two conditions must be satisfied:

1. The product of the sodium and chloride concentrations on one side of the membrane must equal that on the other, i.e.

$$[Na^+]_A [Cl^-]_A = [Na^+]_B [Cl^-]_B$$

2. If electroneutrality is to be maintained, positive and negative charges must balance. The only distribution which satisfies both these requirements is that shown in figure 1.7.

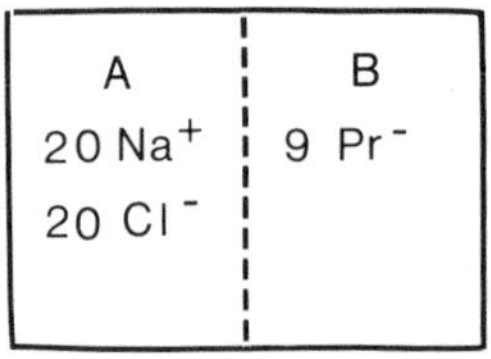

A	B
20 Na$^+$	9 Pr$^-$
20 Cl$^-$	

Figure 1.6   NaCl solution separated from protein solution by a salt-permeable membrane.

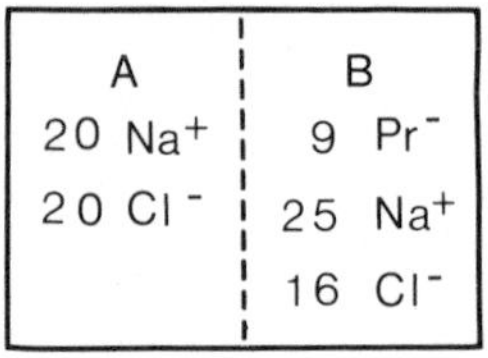

A	B
20 Na$^+$	9 Pr$^-$
20 Cl$^-$	25 Na$^+$
	16 Cl$^-$

Figure 1.7   Distribution of ions in a Donnan equilibrium.

Why should this distribution be stable? Consider what would happen if sodium ions were to diffuse down their concentration gradient from B to A. A positive electrical potential would develop in A, preventing more sodium from entering. Similarly, diffusion of chloride ions from A to B would lead to B becoming electrically negative, preventing more chloride ions from entering. In fact only a minute imbalance in the number of anions relative to the number of cations in a cell will give rise to a large electrical potential. The potential difference developed across the cell membrane will depend on the membrane capacitance, but a very approximate idea of the magnitude involved can be illustrated by reference to a "typical" cell, in which internal potential of the order of -10 mV would result if the number of anions exceeded the number of cations by one part in a million. The interior of living cells (except for nerve cells during the passage of an action potential) is in fact always at a negative electrical potential, usually between -60 and -80 mV, relative to the extracellular fluid bathing it.

The production of a potential as a result of diffusion of ions is also caused by the fact that cell membranes are usually more permeable to potassium ions than to sodium ions; potassium ions therefore diffuse out faster than sodium ions diffuse in. Diffusion of sodium, chloride and potassium ions can usually account for the magnitude of the membrane potential, which can be calculated by applying the *Goldman constant field equation*:

$$E = \frac{RT}{F} \ln \frac{P_{Na}[Na^+]_{in} + P_K[K^+]_{in} + P_{Cl}[Cl^-]_{out}}{P_{Na}[Na^+]_{out} + P_K[K^+]_{out} + P_{Cl}[Cl^-]_{in}}$$

E is the potential difference, R the gas constant, T the temperature in kelvins, F the faraday and P_X the relative permeability of the membrane for the ion X. As an example, in the membrane of the squid giant axon, $P_{Na} : P_K : P_{Cl} :: 0.04 : 1 : 0.45$.

Active transport mechanisms may also give rise to a transmembrane potential, by effecting a net transfer of charge across the membrane; these are known as *electrogenic pumps*. It is often difficult to decide whether a pump is electrogenic or not, because of the other effects giving rise to membrane potentials. For example a pump which exchanges one cation for another in equal numbers can indirectly give rise to a potential difference, if there is a difference in the permeability of the membrane to the two ions, without itself being electrogenic.

Because of these electrical effects we need to consider the *electrochemical gradients* for ions in order to decide whether their distribution is passive or dependent on some active pumping mechanism. If we consider a cell where, for a given ion, A_{in} and A_{out} are the activities inside and outside the cell, F_{in} and F_{out} the fluxes of the ion into and out of the cell and E is the potential difference across the cell membrane, the *Nernst equation* will hold if the system is not subject to an active ion pumping mechanism:

$$\frac{F_{in}}{F_{out}} = \frac{A_{in}}{A_{out}} \cdot e^{zFE/RT}$$

where z is the valency of the ion (positive for cations, negative for anions) and the other symbols are as above.

If the system is in equilibrium, $F_{in} = F_{out}$. Since R and F are constants, the equation can be simplified and expressed in terms of $\log_{10}$ as:

$$\log \frac{A_{in}}{A_{out}} = \frac{E}{58}$$

at 18°C, for monovalent ions. Thus, for an intracellular potential of $-58\,mV$, anions will be in equilibrium when their extracellular fluid concentration is 10 times their concentration in the cell. The low concentrations of chloride and bicarbonate ions in cells are as expected, given the negative intracellular potential, but the high intracellular potassium and low sodium concentrations characteristic of virtually all living cells must be due to an active process.

The sodium pump

It is possible that a small amount of potassium could be bound within the cytoplasm, but this is likely to be of significance only in cells such as freshwater protozoa in which the potassium concentration is very low. It has long been recognised that a mechanism must exist to actively pump potassium ions into cells and sodium ions out. This mechanism has been called the "sodium pump" and it has been studied in a variety of animal cells. One of the most convenient to experiment with is the red blood cell. The mammalian erythrocyte has no nucleus or other organelles, and can be thought of in simple terms as a sac containing a very concentrated haemoglobin solution. If placed in hypotonic solution it will swell and burst, losing its haemoglobin. If burst cells are incubated in isotonic solution at 37°C the ruptured cell membranes will re-seal to form what are known as red cell "ghosts". The composition of the solution in which they are sealed can be varied to produce ghosts with different internal compositions. The ghosts can then be placed in various solutions to study the ion pumping activity of the cell membrane, by techniques such as the use of radioactive isotopes to follow the movement of sodium and potassium ions. Figure 1.8 illustrates what happens in a typical human erythrocyte. ATP is converted to ADP and inorganic phosphate (shown as P in all diagrams—it is in fact a mixture of HPO_4^{2-} and $H_2PO_4^-$ ions at physiological pH) within the cell to supply energy to the pump. Two potassium ions are transported in and three sodium ions out for every molecule of ATP hydrolysed.

The fact that hydrolysis of ATP is an essential step in the pumping action means that an enzyme—an ATPase—is involved. It functions only in the presence of both sodium and potassium ions as illustrated in figure 1.9 so is referred to as $Na^+ + K^+$-activated ATPase. Note the relationship between the K_m for sodium and the internal sodium concentration and the K_m for potassium and the external potassium concentration. (The K_m is the concentration of substrate at which an enzyme-mediated reaction proceeds at half its maximum velocity, and is a measure of the affinity of

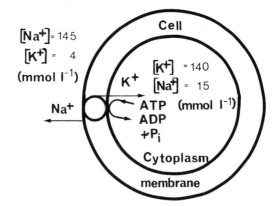

Figure 1.8 The sodium pump in an erythrocyte membrane.

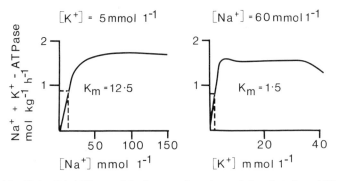

Figure 1.9 $Na^+ + K^+$-ATPase activity (measured as release of phosphate from ATP in mol kg wet $wt^{-1} h^{-1}$) in herring gull salt gland homogenate. (a) Effect of varying the sodium concentration with the potassium concentration kept constant at $5 \, mmol \, l^{-1}$. (b) Effect of varying the potassium concentration with the sodium concentration kept constant at $60 \, mmol \, l^{-1}$. Redrawn and modified from Bonting (1970).

the enzyme for the substrate.) The enzyme is obviously adapted to work efficiently when exposed to internal sodium and external potassium concentrations.

Investigation of the mechanism of the sodium pump has concentrated on the properties of the enzyme, and it is now known that the enzyme molecule itself is in fact the pump. This has recently been confirmed by the incorporation of $Na^+ + K^+$-activated ATPase molecules extracted from cell membranes into artificial lipid membranes, where they function as sodium pumps. The pump is specifically inhibited by the drug *ouabain* which is therefore a very useful indicator of the presence of a $Na^+ + K^+$-activated ATPase pumping mechanism. One ouabain molecule binds per enzyme molecule, probably close to the potassium binding site, since the inhibition produced by low ouabain concentrations can be partly overcome by increasing the potassium concentration of the incubation medium. Cells contain other ATPases, so to determine the $Na^+ + K^+$-activated ATPase activity, control incubations in the absence of either sodium or potassium, or alternatively in the presence of ouabain, must be carried out to determine the basal ATPase activity. This is often referred to as the Mg^{2+}-activated ATPase activity, although $Na^+ + K^+$-activated ATPase is also dependent on the presence of magnesium, since it utilises the magnesium salt of ATP as a substrate. These control values must be subtracted from the activity in the presence of both sodium and potassium and in the absence of ouabain to give the $Na^+ + K^+$-activated ATPase activity.

The molecular mechanism of the sodium pump

$Na^+ + K^+$-activated ATPases have been extracted from a variety of membranes. Each molecule is closely associated with a number of membrane phospholipid molecules—attempts to remove them usually lead to a loss of activity and the detailed structure of the enzyme is not yet known. It has a molecular weight of about 250 000 dalton and probably consists of two 90 000 and two 45 000 sub-units although there may be some variation between species. The small sub-units are glycoproteins, the others are large enough to span a cell membrane.

Many models of the pumping mechanism have been proposed but recent evidence (Karlish, Yates and Glynn, 1978) favours the following theory. Biochemical evidence, using inhibitors which stop the reactions of the $Na^+ + K^+$-activated ATPase at intermediate stages, suggests that the enzyme molecule can exist in two different conformational states which have been called E1 and E2. E1 is stable in combination with ATP and has ion-binding sites facing inwards towards the cytoplasm. These have a higher affinity for sodium ions than for potassium ions so pick up the former from the cytoplasm. This causes release of ADP from the enzyme-sodium-ATP complex, leaving a phosphorylated form of E1 which is unstable and undergoes a configurational change to form E2, in which the ion-binding sites (still with sodium ions attached) face the exterior. E2 has a higher affinity for potassium than for sodium ions, so the latter are released into the extracellular fluid and the former picked up. The binding of potassium ions makes the phosphorylated form of the molecule unstable, so the phosphate is lost. The non-phosphorylated enzyme can now bind ATP, which produces another change in configuration, back to E1, translocating the ion-binding sites (still with potassium ions attached) to a position facing the cell interior. Since E1 has a higher affinity for sodium than for potassium ions, it releases the potassium ions and picks up sodium ions and the process can be repeated. The process is represented diagrammatically in figure 1.10. The ion-binding sites could well be in a channel through the centre of the molecule accessible from both sides so that the actual distance the ions have to be moved is small, but this is only speculation in the absence of detailed structural information.

Such a cyclical model accounts for most of the observed facts except for evidence that sodium and potassium binding sites are occupied simultaneously. This could be reconciled with the model if the pump existed as a dimer, with one unit binding sodium ions whilst the other bound potassium ions.

Even if the sodium and potassium ions bind to different parts of the

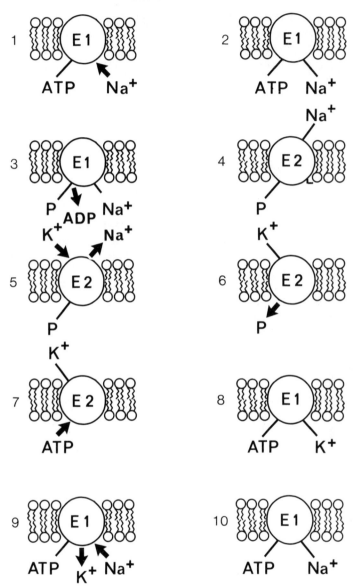

Figure 1.10 Diagrammatic representation of the events taking place during one cycle of the sodium pump. The Na$^+$+K$^+$-ATPase molecule, in its E1 or E2 form, is shown in a phospholipid membrane with the extracellular fluid side upwards and the cytoplasmic side downwards. For details see text.

molecule, how does any biological system differentiate between two such chemically similar molecules? The affinities of a wide range of biological systems for the ions of the alkali metals, lithium, sodium, potassium, rubidium and caesium have been studied. The number of possible affinity sequences is 5!, or 120, but only 11 are found in nature. This exactly parallels the position with respect to ion-sensitive glass electrodes, where the same 11 selectivity series are found out of the possible 120. The glass of ion-sensitive electrodes is made from a mixture of Al_2O_3, SiO_2 and Na_2O. The selectivity series exhibited by a particular mixture depends on the ratio of Na^+ to Al^{3+} ions, as $(AlOSi)^-$ ions are shielded to a greater or lesser extent by Na^+ ions. The strength of the negative charges on the surface of the glass seems to be the factor determining which selectivity series is operative, and the observed findings can be interpreted in terms of the relative electrostatic attraction between the ions and the glass and between the ions and water molecules. For example, if a surface charge is very strong a lithium ion will be preferentially bound, as, once stripped of its large shell of water of hydration, its centre of charge will be close to the surface of the glass. A weak surface charge, on the other hand, will not attract lithium ions, which will not be able to approach closely because of the greater attractive forces of the water of hydration. The smaller hydrated caesium ion, however, will be able to approach closely enough to be attracted.

Selectivity ratios can be calculated on theoretical grounds from the free energies of hydration and the free energies if ion-site interactions, and the same 11 sequences are found. Changing the strength of the negative charges will result in switching from one sequence to another. A similar mechanism may exist in biological systems, although the nature of the negative charge is not known. It could result from acidic groups or dipoles such as exist in $C{=}O$ bonds. The binding of strongly electronegative groups, such as phosphates, to a protein, with the consequent effects on electron distribution through the molecule, could be a means of altering the selectivity of ion binding sites.

The sodium pump and osmoregulation

$Na^+ + K^+$-activated ATPase is rarely found in bacterial and plant cell membranes and does not occupy the same role as it does in animal membranes, as in these organisms accumulation of potassium by cells is generally not linked to sodium extrusion. Instead, electroneutrality is maintained by extrusion of hydrogen ions. $Na^+ + K^+$-activated ATPase activity has been demonstrated in almost all animal cell membranes studied, except in those of protozoa. A few instances of cells lacking the enzyme illustrate its importance in maintaining cellular ionic concentrations. Cat and dog erythrocytes lack $Na^+ + K^+$-activated ATPase, and have high internal sodium and low internal potassium concentrations. Sheep red blood cells are either high or low potassium; the latter suffer from a genetically-determined lack of $Na^+ + K^+$-ATPase.

But is there any osmoregulatory significance in the high potassium and low sodium concentrations of almost all living cells? Life began in sea water, which is mainly a sodium chloride solution, but when cell membranes evolved they must at a very early stage have developed active pumping mechanisms to replace sodium by potassium within the cells. The biochemical processes of most forms of life are obviously adapted to operate most efficiently in a 100 to 200 mmolal potassium solution, as might be expected after thousands of millions of years of evolution in such an environment. But why did they change from the primeval sodium chloride solution in which they originated? One possible explanation is suggested by the situation represented in figure 1.7. The first living cells obviously needed to retain a variety of organic molecules within the cell membrane. If the membranes were permeable to water this would have led to an osmotic influx of water, even if salts were allowed to equilibrate. In fact a Donnan equilibrium would exist, and the intracellular concentration of the permeant ions would have been slightly higher than their external concentration (in the example given in figure 1.7 there are 41 $Na^+ + Cl^-$ ions in the cell and 40 outside). Since cell membranes are always much less permeable to salts than to water, this inequality would have added to the osmotic imbalance.

There are several ways in which cells could have responded to having an osmotic concentration greater than the external medium. They could have evolved cell walls strong enough to resist the hydrostatic pressure created by the osmotic influx of water, up to the point where the hydrostatic pressure balanced the osmotic pressure. They could have evolved mechanisms for continually expelling the water which entered by osmosis. Or they could have actively pumped salt out of the cell to keep the internal concentration lower than the equilibrium value. Plant, bacterial and fungal cells adopted the former solution and it is tempting to speculate that animal cells adopted the latter at a very early stage in evolution, as evidenced by the widespread distribution of the sodium pump. There are two ways in which such a pump will tend to reduce the internal osmotic concentration. Only two molecules of potassium are pumped in for every three molecules of sodium pumped out and, since cell membranes are more permeable to potassium than to sodium ions, the former will diffuse out faster than the latter will diffuse in.

What evidence is there that animal cells osmoregulate in this manner? Regulation of cell volume appears to be an active process since cells swell in the presence of metabolic inhibitors (e.g. dinitrophenol or cyanide). The presence of sodium in the external medium is necessary for cells to maintain a constant volume in iso-osmotic solutions, i.e. an iso-osmotic solution of a salt other than sodium chloride is not isotonic, presumably because in such solutions the internal sodium is quickly depleted and sodium extrusion comes to a halt. But evidence that the sodium pump, as described above, is involved in cell volume regulation is lacking. Ouabain, the potent inhibitor of $Na^+ + K^+$-activated ATPase, does not inhibit the volume-regulatory process, so it has been suggested that some other form of sodium pump is involved.

The details of the cellular osmoregulatory mechanisms are thus far from established, but it is clear that as long as animals have the correct salt solutions bathing their cells they are able to regulate the amount of water and solutes within the cells. With the evolution of multicellular animals, the main osmoregulatory problems have shifted from maintaining the osmotic concentration of the intracellular fluid to maintaining the osmotic concentration of the extracellular fluid, and this will be the subject of most of the rest of the book.

Salt and water movement across epithelia

Various organs are involved in the regulation of extracellular fluid composition in different animals. Salt-transporting epithelia play a vital role, and in fact it is usually more convenient to study ion transport across these tissues rather than across cell membranes. Much of what we know about the movement of salts and water across epithelia, and in particular the mechanisms of hormonal stimulation of these processes, comes from experiments on frog skin and toad bladder, so some of this research will be briefly outlined. All work on salt transport across epithelia owes much to the pioneering work of Ussing on frog skin. He devised an apparatus in which an area of skin was clamped between two Ringer-filled chambers, allowing the transepithelial electrical potential to be measured. This was found to be as high as, or sometimes greater than, 100 mV, serosal side positive. (The serosal surface is the "inside" of the skin or the "outside" of the bladder, i.e. the "blood" side. The external surface of the skin, or the luminal surface of the bladder, is referred to as the mucosal side to avoid confusion.) The potential is due to the active transport of sodium in a mucosal to serosal direction; this can be illustrated by the addition of ouabain (which blocks the sodium pump) to the serosal bath, which reduces the potential to zero.

Sodium transport does not lead to a continuous buildup of positive charge on the serosal side of the skin, because the greater the potential, the greater the tendency for chloride ions to move down their electrochemical gradient and cancel out part of the charge. The potential difference developed depends on the balance between sodium transport and chloride diffusion. Since a flow of charged particles constitutes an electric current, and 1 mol of a monovalent ion carries a charge of 1 faraday, the "sodium current" can be calculated from the isotopically measured net flux (influx − outflux) of sodium. Ussing found that if an electric current was passed to reduce the transepithelial potential to zero this so-called short circuit current (scc) was exactly equivalent to the rate of sodium transport (see chapter 11). This relationship holds for frog skin and toad bladder (but not for many other epithelia) and is the main reason for their popularity in the study of sodium transport; the activity of the sodium pumps can be followed by watching the dial on an ammeter.

Koefoed-Johnson and Ussing (1958) proposed a model for transport through a frog skin cell which is illustrated diagrammatically in figure 1.11. The sodium pump sites are localised on the serosal side of the cells

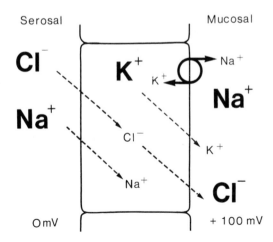

Figure 1.11 Transport through a frog skin cell. Sizes of symbols suggest the relative concentrations of the different ions. Dotted lines represent diffusion down electrochemical gradients. $_{K^+}$ ⟳ $^{Na^+}$ represents the sodium pump.

(ouabain inhibits transport from this side) and keep intracellular concentrations of sodium low and potassium high, as in most cells. Net transport of sodium across the cells occurs because of this asymmetrical distribution of $Na^+ + K^+$-ATPase, and because of differences in permeability characteristics of different regions of the cell membrane. The apical (mucosal) membrane was assumed to be permeable to sodium ions but not to potassium ions, and the basal (serosal) membrane to be permeable to potassium ions but not to sodium ions. Sodium ions would then diffuse from the external solution into the cell down their concentration gradient, to be pumped out across the basal membrane by electrically neutral pumps. The potassium ions pumped into the cell in exchange would tend to diffuse out down their concentration gradient across the basal membrane. The "sodium current" would thus be carried across the apical membrane by sodium ions and across the basal membrane by potassium ions.

This model is obviously a gross oversimplification. Skin consists of several cell layers and several different cell types, but there is evidence that the cells of the outermost living layer, the stratum granulosum (which underlies the keratinised stratum corneum) are responsible for the transport properties. One study, using microelectrodes, showed that the apical membranes of what were presumed to be stratum granulosum cells contributed 74% of the electrical resistance of the epithelium, and that this value rose to 96% after the addition of the drug amiloride, which blocks sodium channels in membranes, to the mucosal bath (Helman and Fisher, 1977). There is however an appreciable conductance through shunt pathways between the cells, and they may constitute the major route for the diffusion of chloride ions (MacKnight et al., 1980).

The sodium pumps are probably located mainly on the lateral cell membranes, pumping sodium into the intercellular spaces, since further cell layers lie beneath the basal membranes. A major problem with the model is its reliance on passive sodium entry across the apical membranes. Krogh demonstrated that frogs could take up salts when kept in only $10\,\mu\mathrm{mol}\,l^{-1}$ sodium chloride solutions. Since the cell interior is positive relative to the mucosal solution, it is difficult to see how the intracellular sodium concentration can be low enough for an electrochemical gradient favouring sodium entry to exist. Measurement of the intracellular sodium level does not help resolve this problem, since not all of it is part of the transport pool. There is good evidence for active secretion of hydrogen ions across the mucosal membranes, and this process may create the necessary electrical gradient to favour sodium entry. Protons are produced during metabolism, and the metabolic rate of frog skin cells depends on the rate of sodium transport. This may be a neat way of balancing the sodium extruded by the baso-lateral pumps and the amount entering across the apical membranes (Ehrenfeld and Garcia Romeu, 1977; 1980). Whatever the coupling mechanism may be, it is clear that anything which increases mucosal entry of sodium, e.g. treatment with the hormone arginine vasotocin (see chapter 6), leads to an increased pumping rate which maintains the low intracellular sodium ion level.

Hormonal stimulation

Studies on salt and water movements in osmoregulatory organs in a variety of animals have demonstrated the importance of endocrine control mechanisms. Two sorts of hormones are involved throughout the vertebrates (with the possible exception of the Agnatha)—neurohypophysial peptides and adrenal steroids. Because so much information has been obtained from studies on frog skin and toad bladder, the mechanism of action of two amphibian hormones, arginine vasotocin and aldosterone, will be considered, although it must be remembered that a wide variety of other hormones are involved in the regulation of osmoregulatory processes in both vertebrates and invertebrates.

Vasotocin is released from the pituitary in response to dehydration and aldosterone is released from the adrenals in response to reduction in extracellular fluid sodium concentrations in Amphibia (chapter 6). The mechanism of vasotocin action on amphibian skin and bladder is complex. It has been studied intensively, largely in the hope of gaining better understanding of the effects of the structurally similar antidiuretic hormone (see chapter 9) on the human kidney tubule, which frog skin and toad bladder are presumed to resemble in some respects. The hormone binds to receptors in the baso-lateral membranes which stimulate the enzyme adenylate cyclase to convert ATP into cyclic AMP. This intracellular messenger initiates several events within the cells, leading to an increase in the permeability of the apical membranes. Permeabilities to both water and sodium ions are increased, but not by a common action since either one can be affected independently of the other. Microtubules and microfilaments evidently play a key role in these processes, as the hormone-induced increase in osmotic permeability can be inhibited by colchicine. Rearrangement of proteins within the apical membrane may involve microtubules or microfilaments, or they may be involved in the fusion of cytoplasmic vesicles with the membrane which occurs during stimulation and which may change the membrane permeability characteristics by the incorporation of vesicle membrane. Direct or indirect (via stimulation of metabolism) stimulatory actions of vasotocin on the sodium pumps have also been suggested.

Vasotocin acts quite rapidly, the effect developing after a lag of only a few minutes. Aldosterone has a longer-lasting action with a long latent period (about 90 min) during which the hormone is transported to the nucleus, where it induces mRNA synthesis leading to the production of several new proteins which, in the case of toad bladder, range in size from 17 000 to 38 000 daltons. These stimulate sodium transport, but in spite of many years' research the mechanism by which they do this is still disputed. They may increase the availability of metabolic substrates, leading to increased ATP production to fuel the sodium pump, or they may act as permeases to increase the permeability of the apical membrane to sodium. Another suggestion is that they increase carbonic anhydrase activity, leading to an increased production of hydrogen ions which facilitate sodium entry. They may of course have more than one site of action.

CHAPTER TWO

LIFE IN THE SEA

TOWARDS THE END OF THE NINETEENTH CENTURY IT WAS DISCOVERED THAT animal extracellular fluids closely resembled sea water in their ionic composition. With the observation that all animal phyla may be found in sea water, yet several are missing in whole or in part from fresh water or land (e.g. echinoderms, pogonophora, cephalopods), this ionic information formed the basis of the generally accepted hypotheses of Haldane, Bernal and Oparin that the origin and early evolution of all living organisms took place in the sea—the environment least hostile to protoplasm. Some of the current ideas about the subsequent penetration of the more demanding littoral, terrestrial, brackish water and freshwater environments are displayed in figures 2.1, 2.2 and 2.3. These figures are greatly simplified, especially in the case of the invertebrates where only a few conspicuous groups are shown. Also, it has to be remembered that our knowledge of the evolutionary history of many groups is patchy, often confused and constantly changing as more fossil evidence becomes available. The examples selected do, however, illustrate the complexity of the changes which have occurred.

The marine environment

The open sea, away from the coasts and freshwater influences, is remarkably stable in its physiochemical conditions. This stability depends partially upon the huge volume of sea water (roughly $12 \times 10^8 \, \text{km}^3$), but also upon the anomalous characteristics of water as a solvent (which derive from its molecular structure and powerful hydrogen bonding). Water has an extremely high heat capacity (3,000 times that of air), thermal conductivity and dissolving power. Even in surface oceanic waters, sea temperature extremes range only from about -1.4 to $+27.5°C$ between polar and equatorial latitudes, whereas comparable air temperature values

25

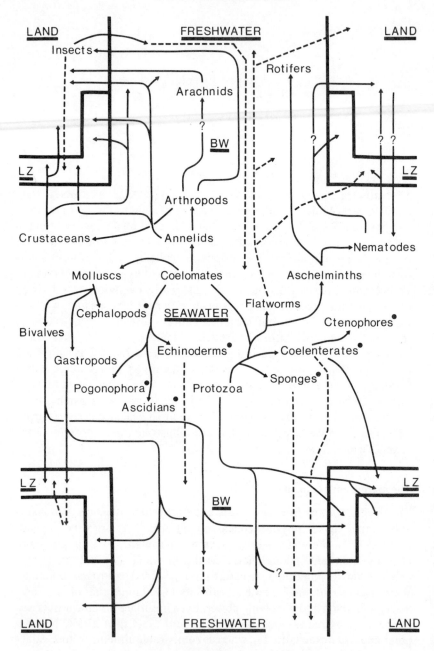

Table 2.1 The constituents of oceanic water of $35\%_{oo}$
salinity (modified from Harvey, 1955)

Ion	$g\,kg^{-1}$	$mmol\,kg^{-1}$
Sodium	10.77	468
Magnesium	1.30	53
Calcium	0.409	10.2
Potassium	0.388	9.9
Chloride	19.37	546
Sulphate	2.71	28
Bicarbonate	0.14	2.3
Bromide	0.065	0.8
Boric acid (as H_3BO_3)	0.026	0.4

are $-68.5°C$ and $+58°C$. At depth in the oceans, temperature is virtually constant at about $+2°C$ to $+3°C$ in tropical, subtropical and temperate regions; only close to the poles does it become slightly colder (0 to $+1.5°C$). Chemical stability is even more remarkable; the salinity, which represents the sum of the dissolved inorganic salts in sea water, varies very little over the oceans of the world. The range is from 32 to 38 parts per thousand $(\%_{oo})$. Values outside this range are found only in enclosed seas such as the brackish Baltic (chapter 3) or the Red Sea which, with little freshwater input and intense sun-driven evaporation, has a salinity of more than $40\%_{oo}$. Again, deep water is characterised by even greater stability, with salinities always falling between 34.5 and $35\%_{oo}$. If salinities are relatively constant in the oceans, the proportions of salts which make up those salinities are even more stable. Estimates of salinity in samples of sea water were performed for many years (before the development in the 1960s of conductivity meters) by simply measuring the chloride content, since as the ionic ratios of sea water were so constant the total salt content could be calculated from it with great precision. Although it is probably true that all the elements in the periodic table are present in sea water, many occur only at vanishingly small concentrations, and the nine ions whose concentrations are given in Table 2.1 make up about 99.5% of the total salt content of sea water.

Figure 2.1 Outline of the radiation of invertebrate groups from sea water into other habitats. BW = brackish water, LZ = littoral (i.e. intertidal) zone, ● = groups which are exclusively or almost exclusively marine in distribution. Dotted lines indicate penetration by very few species, ? = uncertainty regarding route by which species reached their present habitat. N.B. Phylogenetic relationships are controversial; the scheme shown is only one of those which might be constructed.

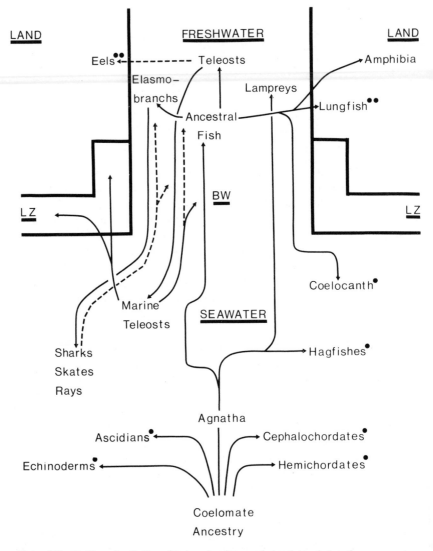

Figure 2.2 Outline of radiation of lower chordates and chordate relatives from sea water into other habitats. Symbols as for figure 2.1 except for ●● which indicates species which are terrestrial for short periods only.

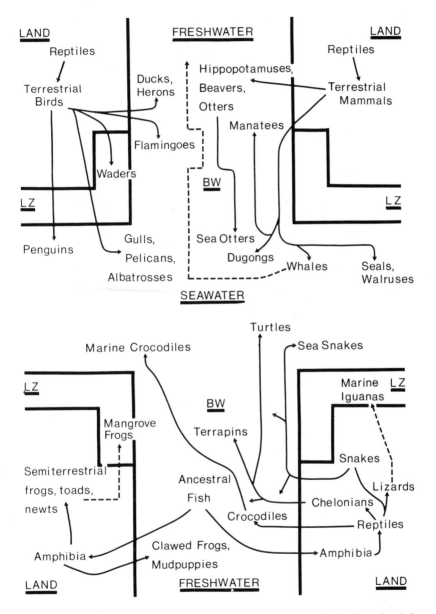

Figure 2.3 Outline of radiation of higher vertebrates from land into other habitats. Symbols as for figures 2.1 and 2.2.

The ionic composition of marine invertebrates

In view of the ocean's huge volume, it is obvious that any significant changes in concentration or proportion of salts in sea water would take long periods to accomplish. Isotope determinations show that the age of deep water may be reckoned in centuries, or even thousands of years, despite the mixing processes of the seas. In the early days of osmotic and ionic studies, when it was found that marine animals' body fluids deviated somewhat in their ionic ratios from the surrounding sea water (see Table 2.2), it was suggested that the composition of sea water had changed often in the past, and that the ionic ratios of body fluids reflected the composition of sea water at the time of each species' evolution. As more animals were studied, and all showed some ionic deficits or surpluses by comparison with the sea, yet also differed from each other, this idea became patently absurd. In any case, the geochemical evidence available at present suggests that the composition of salts in sea water has changed little during geological time. On the other hand the evidence concerning overall salinity levels in the past is somewhat conflicting, and it remains possible that the sea has been at times rather more dilute or more concentrated than it is at present. However, such changes of concentration would

Table 2.2 Extracellular fluid ionic concentrations in some marine invertebrates (from Robertson, 1957) and in *Myxine* (from Bellamy and Chester Jones, 1961)

Group	Concentrations of ions as percentage of seawater ionic concentrations					
	Na^+	K^+	Ca^{2+}	Mg^{2+}	Cl^-	SO_4^{2-}
(A) Coelenterates						
Aurelia aurita	99	106	96	97	104	47
(B) Echinoderms						
Marthasterias glacialis	100	111	101	98	101	100
(C) Tunicates						
Salpha maxima	100	113	96	95	102	65
(D) Annelids						
Arenicola marina	100	104	100	100	100	92
(E) Crustaceans						
Maia squinado	100	125	122	81	102	66
Carcinus maenas	110	118	108	34	104	61
Nephrops norvegicus	113	77	124	17	99	69
(F) Molluscs						
Pecten maximus	100	130	103	97	100	97
Sepia officinalis	93	205	91	98	105	22
(G) Vertebrates						
Myxine glutinosa	117	91	61	38	102	—

have taken millions of years to complete, and would have allowed plenty of time for species to acclimatise.

Physiologists interested in osmotic phenomena have paid relatively little attention to stenohaline (i.e. intolerant of salinity change) seawater animals of exclusively marine ancestry (almost all invertebrates but including the most primitive vertebrates, the hagfishes). This is mainly because there are more challenging problems in the study of the adaptations of their more euryhaline (i.e. tolerant of significant salinity changes) brackish-water or intertidal relatives—as might be expected scientists tend to select the more "interesting" species for their studies! However, all primarily marine animals studied have body fluids which are iso-osmotic with the surrounding sea water and are therefore not exposed to osmotic stress. Despite this equable situation, reference to Table 2.2 shows that in all cases there are discrepancies between sea water and body fluid concentrations of certain ions. The existence of a Donnan equilibrium, due to the presence of protein in the body fluids, means that some differences would be expected, but this factor does not account for the degree of divergence commonly found. In particular there is a general tendency for extracellular fluid sulphate levels to be significantly lower than in sea water. Crustacea tend to have low haemolymph magnesium concentration while several animals accumulate potassium to some extent. These ionic imbalances have certain consequences. Since some ions are in deficit with respect to sea water they must be excreted, either by active transport or by iso-osmotic removal in urine passed out of the body via kidneys or analogous structures. The secretion of urine results in a loss of bulk fluid, so volume-regulatory mechanisms are required by the animal concerned. Those ions which exist in surplus within the body fluids demand active uptake mechanisms and sites. All these features of stenohaline marine animals have served as preadaptations for the colonisation of more demanding habitats such as the intertidal zones and estuaries.

Before leaving the topic of ionic imbalances between body fluids and sea water, a few special features exhibited by some organisms should be mentioned. From Table 2.2 it may be seen that echinoderms like *Marthasterias* have body fluids nearly identical in ionic composition with sea water. However, even in these animals it has been found that there are localised high coelomic fluid potassium concentrations in the regions of the ambulacral grooves between the tube feet. These heightened concentrations are believed to be associated with muscular activity. Another group of exclusively marine animals are the cephalopod molluscs. Their failure to invade more demanding environments is something of a mystery since, as will be discussed later (in chapter 10), they possess sophisticated ionic and osmotic regulatory mechanisms involved in the control of their buoyancy. Finally there is an interesting story concerning the blood magnesium levels of prawns and crabs which was elucidated by Robertson (1953). He observed (see Table 2.2) that all of the crabs and prawns studied had reduced haemolymph magnesium levels, but that active species such as prawns and portunid crabs had much lower magnesium concentrations than did lobsters or such inactive crabs as *Maia*, the spider crab. Robertson was later able to show that neuromuscular transmission in crustaceans was strongly influenced by magnesium; high concentrations were depressive in effect while low concentrations were stimulatory. Unfortunately this elegant correlation between low blood magnesium, fast neuromuscular transmission, and active behaviour appears to apply only to crustaceans. Reference again to Table 2.2 shows that the cuttlefish *Sepia officinalis*, although one of the quickest swimmers in the sea, has a blood magnesium level almost identical with sea water.

Although the extracellular fluids may be similar in their composition to sea water, the intracellular fluid composition of all marine animals is markedly different, having the high potassium concentrations characteristic of all cells and much lower sodium and chloride levels. Organic solutes make up the balance of the intracellular osmotic concentrations. There is considerable variation in the nature of these intracellular organic osmotic effectors. Amino acids play a very important role in all species studied, but other compounds such as trimethylamine oxide (TMAO), betaine, organic phosphates and other organic acids may be involved. Figure 2.4 shows the main osmotic effectors in the hagfish, *Myxine glutinosa*. The importance of intracellular osmotic effectors such as amino acids to the survival of animals in brackish water habitats will be discussed in chapter 3.

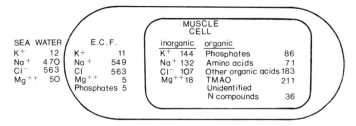

Figure 2.4   Solute concentrations (mmol kg water $^{-1}$) in the serum and muscle intracellular fluid of the hagfish, *Myxine glutinosa*, compared with those of the sea water in which they were kept (data from Bellamy and Chester Jones, 1961).

Hagfish—structure and function of a primitive vertebrate kidney

To end this consideration of primarily marine animals an unusual group of animals will be briefly mentioned—the hagfishes. In some respects the hagfishes are very specialised animals but they have attracted much study because they undoubtedly possess the most primitive of all vertebrate kidneys, having 15 to 20 pairs of segmentally arranged glomeruli. Blood magnesium and sulphate levels are low in *Myxine*, the most closely studied of the hagfishes, and it has been suggested that the kidney is responsible for keeping the plasma concentrations of these ions low. However, magnesium ions are concentrated to a much greater extent in the bile and this may be the main route of magnesium excretion. Hagfishes are also

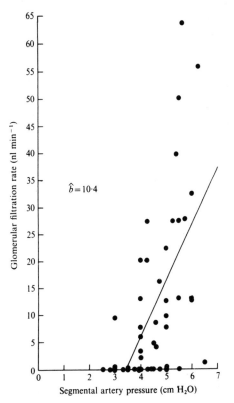

Figure 2.5 Relationship between filtration rate in a single glomerulus (see chapter 4 for information on the measurement of glomerular filtration rate) and pressure in a segmental artery, a branch of which supplies the glomerulus (from Riegel, 1978).

able to secrete quite enormous quantities of slime from the skin and it has been suggested that magnesium is excreted in this copious mucus. Wherever magnesium and other unwanted ions are lost they are excreted as an iso-osmotic fluid, thus causing a volume-regulation problem. Volume regulation (rather than ionic regulation) may be the principal function of the hagfish kidney (and other primitive excretory organs). If hagfish plasma is very slightly hyperosmotic to sea water, possibly because of the small osmotic pressure exerted by low concentrations of organic macromolecules in the plasma, the animals could obtain water from the sea by osmosis. Mechanisms would therefore be necessary to remove any excess. The hagfish kidney at least has the potential to remove a water load since perfusion of individual glomeruli of *Myxine* shows that filtration rate (i.e. primary urine production) rises steeply with increasing arterial pressure (see figure 2.5). Since an expansion of blood volume as a result of osmotic inflow of water would tend to cause a rise in blood pressure, it would appear that an automatic volume-regulatory mechanism controlled by blood pressure acting at the glomerulus exists in hagfishes. It has to be admitted that much of this hypothesis is pure speculation, but on the other hand it is known that the Pacific hagfish *Eptatretus stoutii* is capable of removing a water load, since it can maintain constant volume in 80% sea water.

CHAPTER THREE

THE BRACKISH-WATER ENVIRONMENTS

RELATIVELY FEW SPECIES ARE CHARACTERISTIC SOLELY OF BRACKISH water;* the great majority of animals which have colonised the environments listed below are euryhaline marine forms. Freshwater species are not normally found in waters of more than 100 to 200 mOsm l^{-1}. On the other hand, freshwater animals are descended from marine ancestors which must have gradually evolved the mechanisms of hyperosmotic regulation possessed by all freshwater forms (see chapter 4) whilst inhabiting the intermediate environment of brackish water. In the course of evolution many freshwater groups have re-invaded the seas, again passing through brackish water in the process. Some migratory species re-trace these movements in the course of their life cycles; these will be discussed in chapter 5. The brackish-water environments have played a crucial role in the emergence of animal life from its original home, the sea, into the freshwater and terrestrial environments.

A distinction can be drawn between the problems encountered by animals living in stable salinity regions such as may be found in the Baltic Sea, the White Sea or Chesapeake Bay and those found in areas of varying salinities. Many species living in, for example, the Baltic Sea encounter little change in environmental salinity throughout their life history since although osmolarities of from 50 to 900 mOsm l^{-1} are found, concentrations at any one point change little, even seasonally, so that osmoconformers are under little or no osmotic stress. On the other hand, the

* The osmolarity of the open sea varies from about 900 to about 1200 mOsm l^{-1}. The term "brackish water" has been applied to all waters having a salt concentration lower or higher than sea water, but it has been suggested that it should be confined to waters containing more dissolved salts than fresh water but less than sea water, i.e. in the range from 15 to 900 mOsm l^{-1}. Media ranging from 1200 to 2300 mOsm l^{-1} can then be referred to as "hypersaline waters" and those above 2300 mOsm l^{-1} as "brine waters". These will be discussed later, but most of this chapter will be devoted to organisms inhabiting brackish water as defined here.

body fluids of animals exposed to fluctuating-salinity environments can often be very hyperosmotic or hypo-osmotic to the external medium; problems of water and salt regulation therefore arise.

Types of fluctuating-salinity environment

1. The littoral zone

The littoral zone, which is uncovered on each tide, is often subject to considerable freshwater influence, resulting in tidal salinity fluctuations. At low tide, shallow pools can be rapidly diluted by rainfall or concentrated by evaporation. Animals living at or near the surface of flat sandy beaches can be exposed to the influence of almost pure fresh water during violent storms. These effects are often most pronounced in tropical areas. In temperate regions freshwater run-off is usually in the form of small streams, which may be thought of as miniature estuaries (see below), affecting a relatively small portion of the shore. At high latitudes, however, melt water may flow over virtually all of the shore in warm spells during the late spring, thus exposing a high proportion of the fauna to hypo-osmotic conditions at low tide. Strangely enough, pools in the Arctic littoral zone can become extremely saline. During cold weather, when the air temperature is perhaps $-10°C$ or lower, and the sea water much warmer (around $+4$ to $+5°C$), the ebbing tide will leave behind pools which

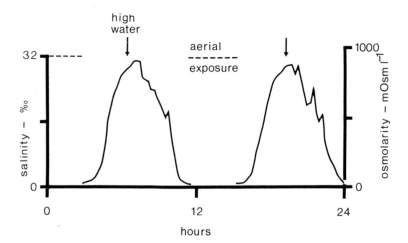

Figure 3.1   Salinity changes during the tidal cycle at a point on the shore of the Conwy Estuary, North Wales (redrawn from Cawthorne, 1979).

rapidly freeze over. When this occurs, the initial ice formed is largely composed of fresh water frozen out of solution, and the remaining unfrozen sea water beneath the ice rapidly becomes concentrated, reaching perhaps 1800 to 2000 mOsm l^{-1} within a few hours.

2. Estuaries

Estuaries, the tidal portions of river mouths, are characterised by salinity fluctuations which have a tidal periodicity (approximately 12.4 h). The amplitude of the fluctuations is most marked where marine and freshwater influences are roughly equal, and decreases as the boundaries of the estuary are approached. The maximum salinity reached generally decreases with distance from the sea. The form and magnitude of the salinity fluctuations are extremely variable and are influenced by tidal state, weather and topography. Figure 3.1 illustrates the sort of short term extreme salinity changes which can occur on the shore of an estuary.

3. Fjords

Fjords are long enclosed arms of the sea and the bulk of the water in them has the same salinity as the open sea. At high latitudes, however, layers of low-salinity water form on the surface of the fjords when snow and ice melt in spring. Benthic organisms on the shores of the fjords are exposed alternately to high and low salinity water as the tide rises and falls.

4. Lagoons

Lagoons are bodies of water, usually shallow, which are in contact with the sea on some high tides. In temperate zones they are often characterised by falling salinity levels during the period of separation from the sea because of rain and freshwater run-off. In semitropical and tropical regions, however, isolated lagoons may become very saline because of evaporation. This type of environment therefore features relatively long periods of gradual salinity change which alternate with abrupt changes in salinity when the lagoon is flushed out by the incoming tide (sometimes at intervals of several months if connection with the sea occurs only on extremely high tides).

5. Mangrove swamps

Mangal communities, in which mangrove trees make up only part of the flora, occur in coastal and estuarine intertidal areas and in lagoons. The salinity conditions they are exposed to thus reflect their particular siting.

6. Salt marshes

Salt marshes are found close to the coast and around estuaries and are low-lying land areas occasionally exposed to sea water or brackish water on very high tides. Free water is usually present as pools or puddles, often of very small size—gammarid amphipods have been found in water lying in cattle hoof prints in such marshes! Because the bodies of water are so small, evaporation or rainfall can produce rapid salinity changes and make this environment probably the most unpredictable and demanding of all brackish-water areas, in which only the most euryhaline species can survive.

Osmoregulators and osmoconformers

The way in which internal osmotic and ionic concentrations are regulated varies in different groups of brackish water animals. Most vertebrates,

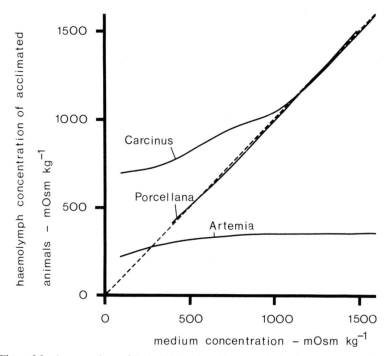

Figure 3.2 A comparison of body fluid and ambient osmolarity for an osmoconformer (*Porcellana*) and for two species showing different types of osmoregulatory response (*Carcinus* and *Artemia*). Broken line is the line of iso-osmocity.

whether of freshwater or terrestrial origin, maintain blood osmolarities of around $300 \, \text{mOsm} \, l^{-1}$ (with some important exceptions which will be discussed below). Different osmoregulatory mechanisms are employed in hyperosmotic and hypo-osmotic external media; these will be discussed in the following two chapters. Marine invertebrates with an internal osmolarity close to that of sea water (about $1000 \, \text{mOsm} \, l^{-1}$) have given rise to the great majority of the fauna of brackish water. In hypo-osmotic media they tend to gain water by osmosis and to lose salts by diffusion; lowered blood concentrations result. It should be noted, however, that in a fluctuating salinity environment there are likely to be occasions when, for some species, osmotic water loss and inward salt diffusion become major problems.

Brackish water animals can be divided into two categories on the basis of their physiological responses to reduced or fluctuating salinities; osmoconformers and osmoregulators. It should be stressed that most brackish-water animals are euryhaline osmoconformers (Kinne, 1971). As far as is known, all echinoderms and coelenterates are osmoconformers as are the majority of polychaetes, bivalve molluscs, cirripede crustaceans and many malacostracan crustaceans. If the environmental salinity is reduced, these animals absorb water and lose salts until their bodies are iso-osmotic with the external medium (figure 3.2).

Volume regulation

Because water movement is faster than salt diffusion, osmoconformers gain a considerable amount of water before osmotic equilibration is attained, causing them to increase in weight and swell up during the period immediately following a reduction in external salinity (figure 3.3). This swelling is disadvantageous and rapidly impairs body activities, including locomotory and food collection mechanisms. Euryhaline osmoconformers are able to eliminate the excess fluid by producing an iso-osmotic urine at an appropriate rate; this ability distinguishes them from stenohaline osmoconformers which are confined to the fully marine habitat. The latter (which includes large groups such as the cephalopod molluscs and most echinoderm species, as well as representatives from most other invertebrate groups) simply swell up and remain swollen if placed in diluted sea water. Figure 3.4 illustrates an interesting example of this difference; young hermit crabs, *Pagarus bernhardus*, living in the shells of *Littorina* and *Nucella* are moderately euryhaline and are common between the tide marks, whereas the much larger older animals which live in *Buccinum*

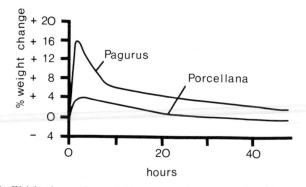

Figure 3.3 Weight changes in euryhaline osmoconformers transferred from a medium of $1000 \, mOsm \, l^{-1}$ to $630 \, mOsm \, l^{-1}$ at time 0 (after Davenport, 1972*b*).

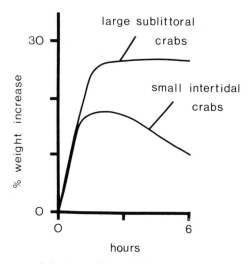

Figure 3.4 Volume regulation in small (euryhaline) and large (stenohaline) hermit crabs, *Pagurus bernhardus*; weight changes after transfer from $1000 \, mOsm \, l^{-1}$ to $630 \, mOsm \, l^{-1}$ (from Davenport, 1972*c*).

shells are sublittoral and stenohaline. Whilst the younger animals can reverse osmotic swelling by increasing urine output, the older crabs cannot produce enough urine to do so. If the nephropores of young crabs are blocked they respond in the same way as do older animals.

The increased urine flow characteristic of euryhaline osmoconformers following dilution of the external medium is, of course, a necessary pre-adaptation for osmoregulation. Osmoregulators face the same problems of volume regulation when exposed to a hypo-osmotic environment and respond in the same way. For example, when the shore crab, *Carcinus maenas*, is transferred from sea water to 50% sea water the hydrostatic pressure in the haemolymph starts to rise as water is absorbed. If the nephropores are blocked the pressure rises steadily, but normally urine is produced by the antennal glands to limit the rise in internal pressure. The time course of such a diuretic response is shown in figure 3.5. Experiments in which haemolymph pressure was raised artificially suggested that the pressure rise was not sufficient, in itself, to explain the increased urine flow. *Carcinus* may be able to detect the change in external osmolarity and regulate its urine production accordingly (Norfolk, 1978). However, large increases in internal pressure produce increases in urine flow, and this may be the mechanism operating in euryhaline osmoconformers (cf. *Myxine*, chapter 2).

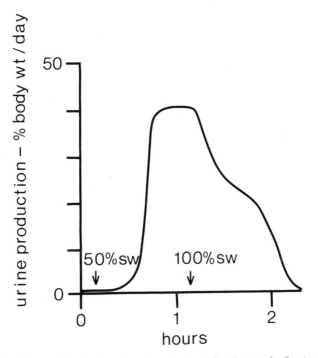

Figure 3.5 Urine production from the antennal gland of a shore crab, *Carcinus maenas*, transferred from 100% sea water to 50% sea water and back (redrawn and modified from Norfolk, 1978).

Intracellular regulation

Allowing the extracellular fluid osmotic pressure to fall in line with that of the environment, which always occurs to some extent, even in osmo-regulators, means that cellular swelling occurs as a result of osmotic water inflow. This would eventually lead to disruption of the cells, but even before such gross damage is done the changes in cell size and alteration in internal geometry are likely to disturb metabolic functions. Diffusion of salts through cell membranes will obviously tend to reduce the osmotic gradient but much of the intracellular osmotic pressure is due to impermeant low molecular weight organic compounds, which exert a so-called "colloid osmotic pressure". Over the last 25 years or so it has been shown that a wide range of osmoconformers react to exposure to low salinities by reducing the colloid osmotic pressure of their tissues, usually by lowering amino acid concentrations. It should be noted, however, that the molecular weights of amino acids, although high enough to prevent diffusion out of the cell, are lower than those associated with truly colloidal substances. The number of amino acids which are regulated varies; for example in *Arenicola* only alanine and glycine are of significance, whereas in other species several amino acids may be involved.

It appears that the amino acids employed in cellular volume regulation are overwhelmingly the so-called "non-essential" amino acids. The mitten

Table 3.1 Amino acid concentrations in the tissues of *Eriocheir sinensis* (data from Bricteux-Grégoire *et al.*, 1962)

Amino acid	Concentrations ($\mu mol/100$ mg wet weight)		% increase
	in fresh water	in sea water	
Alanine	1.39	3.37	142.5
Arginine	2.99	4.13	107.5
Aspartic acid	0.29	0.86	196.6
Glutamic acid	0.84	2.11	151.2
Glycine	4.64	8.00	72.4
Isoleucine	0.08	0.24	200.0
Leucine	0.14	0.40	185.7
Lysine	1.16	1.38	18.9
Phenylalanine	0	trace	—
Proline	0.77	3.50	354.5
Serine	0.21	0.47	123.8
Taurine	1.67	2.06	23.3
Threonine	0.36	1.14	216.7
Tyrosine	0	trace	—
Valine	0	0.5	—
Total	14.54	28.16	93.7

crab *Eriocheir* has been much studied, since it can survive both in fresh water and in sea water. Although an osmoregulator, the species still exhibits great haemolymph concentration changes when migrating between these media and there are concomitant alterations in tissue amino acid concentration to allow cellular volume regulation (Table 3.1). It is clear that alanine, arginine, aspartic and glutamic acid, glycine and proline are the main osmotic effectors. When the mitten crab migrates into fresh water the tissue amino acid levels fall; when it returns to the sea again to breed, the levels rise to prevent cell shrinkage.

The mechanisms involved in the regulation of intracellular amino acid concentrations have received a great deal of attention from biochemists, notably Gilles, Florkin and Schoffeniels. A massive review of the field has recently been published (Gilles, 1979). Much controversy has arisen over the years concerning the means by which intracellular amino acid concentrations are lowered in response to hypo-osmotic stress (low salinity exposure), and of the fate of the lost amino acids. At one time it was suggested that the lost amino acids were incorporated into proteins and thus became osmotically inactive. This is an attractive idea, as it would allow an animal to regulate its cell volume without loss of valuable substrate material, but the available evidence does not support this hypothesis.

When an animal is transferred from sea water to a lower-salinity medium, and therefore exposed to a hypo-osmotic stress, the following sequence of events appears to take place. Firstly, perhaps as a result of osmotic swelling (although this is by no means certain), there is an increase in cell membrane permeability to at least some amino acids (proline, tyrosine, phenylalanine, leucine, isoleucine and valine appear to be regulated primarily by membrane permeability changes) which therefore diffuse into the blood down a concentration gradient. In most cases the extruded amino acids are broken down elsewhere in the body by oxidases, and the nitrogen excreted as ammonia. In *Eriocheir* for instance, it would appear that extruded proline is transported to the three posterior pairs of gills of the crab where it is degraded by the action of "proline oxidase" and the resultant ammonia diffuses easily to the external medium. Secondly, amino acids not regulated by permeability changes may be broken down by the action of oxidising enzymes within the cell itself, the breakdown products being able to diffuse out into the blood. Finally, the amount of production of non-essential amino acids within the cell is reduced. Control of amino acid metabolism within the cell appears to be directly controlled by the prevailing intracellular ionic concentrations. Schoffeniels and Gilles

(1970) have shown that the activities of many enzymes involved in amino acid synthesis and breakdown are profoundly altered by changes in the NaCl content of the incubation medium; this is especially true of glutamate dehydrogenase whose activity is practically halved by a decrease from 400 to 50 mmol l^{-1} NaCl. In consequence, as salts diffuse out of the cells in response to hypo-osmotic stress, the glutamic acid equilibrium will shift in a catabolic direction. It should be noted, however, that the situation is not as clear-cut for other amino acids, and that continuing research is building up a complicated biochemical picture, much involved with the chemistry of anaerobic metabolism. There appears to be little or no evidence for any hormonal control of intracellular amino acid concentrations.

When an animal acclimatised to a dilute medium is returned to sea water, a reverse series of events takes place. Decreased amino acid catabolism, increased synthesis and reduced membrane permeability all conspire to raise intracellular amino acid levels (and hence increase osmotic concentration). It is also possible that amino acids may be actively taken up from the blood, though some of the available information suggests that all amino acids involved in intracellular osmotic regulation are derived from substrates inside the cell.

Regulation which does not involve amino acids also exists; a marine arachnid, the king crab *Limulus polyphenus*, regulates its intracellular osmolarity by changes in the concentrations of low molecular weight nitrogenous compounds of unknown composition.

Some workers have claimed that the reduction in colloid osmotic pressure occurs rapidly enough to prevent cellular swelling in dilute media, but it must be emphasised that much of the data collected to demonstrate the connection between lowered salinity and reduced tissue amino acid levels has been obtained either by collection of specimens of one species from waters of different salinities within stable brackish-water areas, or by experiments involving long periods of acclimatisation to lowered salinities. Recent evidence suggests that, for some species at least, the reduction in amino acid concentrations in response to reduced external osmolarity is of little importance in environments subject to short-term salinity fluctuations as it may take days, or even weeks, to be completed. For example, whereas in the Baltic Sea the mussel, *Mytilus edulis*, shows a linear relationship between external osmolarity and tissue amino acid concentration, amino acid concentrations vary very little in mussels exposed to simulated tidal fluctuations (figure 3.6) although the mean level is lower than in animals taken from full sea water.

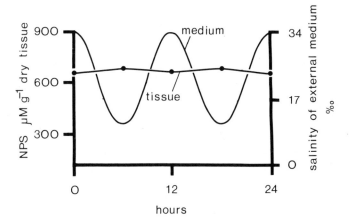

Figure 3.6 Tissue amino acid concentrations (measured as ninhydrin positive substances, N.P.S.) in mussels, *Mytilus edulis*, exposed to fluctuating salinities (redrawn from Shumway *et al.*, 1977).

In environments such as estuaries the periods of exposure to low salinity are relatively short, and there is a tendency (especially in larger animals) for body fluid concentrations to change more slowly than external salt concentrations. Iso-osmocity with the external medium is not therefore attained at low salinities and osmoconformers probably simply tolerate a degree of cellular swelling. In areas where the osmolarity is low over long periods (e.g. lagoons), the amino acid response is certainly important, and in stable brackish water areas it is probably the major factor allowing the gradual colonisation of low salinity regions.

Behavioural osmotic control

For many brackish-water osmoconformers the basic physiological tolerances of changes in external osmolarity are supplemented by behavioural mechanisms which allow animals to avoid exposure to salinities which would exceed these tolerances. These mechanisms take many forms and may be extremely efficient, although they are obviously only of use to animals living in areas of fluctuating salinity.

The lugworm, *Arenicola marina*, is very common in estuaries, living in burrows in sand or mud. Exchange of water between muddy and sandy substrates and the overlying water column is generally poor in estuaries. During periods when freshwater influence is at a maximum, the interstitial water tends to be much more saline than the water passing over the substrate. When the salinity of the overlying water is high, *Arenicola* irrigates its burrow regularly by pumping vigorously from the water column and also, to a certain extent, from the substrate surface layers. The worm spends a great deal of time quite close to the surface. When the overlying osmolarity falls to about $600 \, \text{mOsm} \, l^{-1}$ the worm ceases irrigation and compresses itself at the bottom of the burrow. It rises to the head of the burrow at intervals and pumps water briefly; this has been interpreted as "testing behaviour". If the salinity is too low it quickly stops pumping and returns to the bottom of the burrow. When quiescent in the burrow the worm is subject only to the saline influence of the interstitial

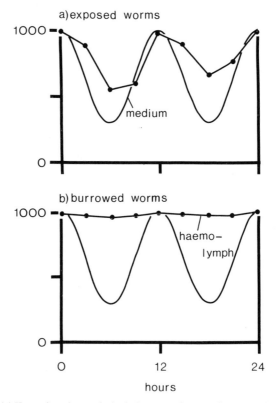

Figure 3.7   (a) Haemolymph osmolarity in lugworms, *Arenicola marina*, directly exposed to conditions of fluctuating salinity. (b) Haemolymph osmolarity of *Arenicola* burrowed in sands beneath a water column of fluctuating salinity (redrawn from Shumway and Davenport, 1977).

water of the substrate, and is thus isolated from the hypo-osmotic overlying water column. The regular testing behaviour enables *Arenicola* to take maximum advantage of the high tide periods when it pumps high-salinity water through its burrow for feeding, respiratory and excretory purposes. This behavioural osmotic control is remarkably effective, as can be seen from figure 3.7. Worms directly exposed to water of fluctuating salinity show substantial changes in body fluid osmolarity and tissue water content; worms burrowed in sand beneath fluctuating salinity water show negligible changes in haemolymph concentration or tissue hydration. Many other estuarine organisms from a variety of phyla burrow into the substrate and are likely to derive similar benefits.

An analogous behaviour pattern is shown by several epibenthic organisms, notably balanomorph barnacles, bivalves, gastropods and hermit crabs. These are all capable of isolating themselves from surrounding low-salinity water; barnacles by closing their scuto-tergal valves, bivalves by adducting their shell valves and gastropods and hermit crabs by retreating into their shells. Limpets and estuarine chitons clamp themselves down onto the rocky substrate when external salinities are low. Such behavioural osmotic control may be as effective as physiological osmoregulation. In figure 3.8 it may be seen that shell valve adduction in *Mytilus edulis* allows the sea water in the mantle cavity to remain above $600 \, \text{mOsm} \, l^{-1}$ in the presence of external osmolarities as low as $3 \, \text{mOsm} \, l^{-1}$.

A penalty of isolation behaviour (including burrowing) is that organisms are cut off from an external food and oxygen supply for part of each tidal cycle. The estuarine distribution of these organisms is likely to be limited not directly by salinity but by the duration of the periods during

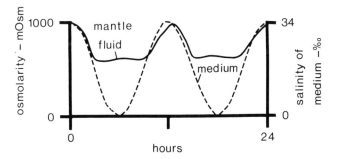

Figure 3.8 Changes in mantle fluid osmolarity in *Mytilus* exposed to fluctuating salinity (redrawn from Shumway, 1977*a*).

which the salinity is high enough to allow feeding and gaseous exchange. Osmoregulators obviously do not suffer from this disadvantage.

Mytilus edulis exhibits a modification of the isolation behaviour pattern which partially offsets the respiratory problem. The mussel's initial reaction to falling external osmolarity is not complete shell valve closure, but closure of the exhalent siphons. This stops effective irrigation of the mantle cavity, and water exchange between the gaping shell valves is relatively slow, but a certain amount of gaseous exchange is still possible. Because of this, and the effectively larger volume of the mantle cavity, the oxygen tension of the mantle fluid declines at a significantly lower rate than it would if the mussels suddenly closed completely at a critical external osmolarity. Eventually the mantle fluid osmolarity does decline and the shell valves are then adducted tightly. However, calculations suggest that this intermediate partial isolation behaviour may gain the mussel as much as one hour's extra oxygen supply per day in an estuarine situation — a worthwhile bonus.

Several organisms, particularly crustaceans, have evolved behavioural responses which keep them away from areas of dangerously low salinity and they therefore avoid the penalties incurred as a result of the isolation responses described above. Most of these species are osmoregulators but a few are osmoconformers. The porcelain crab, *Porcellana platycheles*, is an example of the latter. This anomuran crab is found in the littoral zone of sheltered rocky shores and also occurs in the outer portions of estuaries. It lives in shallow pools at mid-tide level and is therefore likely to encounter lowered salinities. It can only survive indefinitely in water of $500 \, \text{mOsm} \, 1^{-1}$ and above, but can tolerate osmolarities as low as $100 \, \text{mOsm} \, 1^{-1}$ for at least 2 to 3 hours. The crab has three behavioural mechanisms which allow it to avoid prolonged exposure to dangerously low salinities. Firstly, when immersed, it can detect salinity gradients and move along them to areas of higher concentration. The receptor sites involved seem to be on the antennules. Secondly, the crabs can recognise a dangerous external salinity and climb upwards to get out of the water. Finally, they can use their second and third walking limbs to test the salinity of pools before immersing themselves.

It may be seen then that osmoconformers, particularly those with supplementary behavioural osmotic control, can survive in a wide range of salinities and can compete effectively with osmoregulators. Although *Mytilus edulis* has no physiological control over the osmotic pressures of its body fluids, its shell-closure mechanism enables it to maintain as high a haemolymph osmolarity in a fluctuating-salinity regime as the powerful

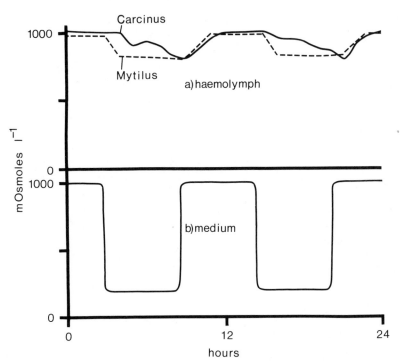

Figure 3.9 Changes in haemolymph osmolarity in *Mytilus* and *Carcinus* exposed to an abruptly fluctuating salinity regime (unpublished data for *Carcinus* and redrawn data of Shumway, 1977*a*).

osmoregulator *Carcinus maenas* (figure 3.9). Indeed, in environments which feature abrupt salinity changes it may suffer rather less haemolymph dilution.

Osmoregulators

As described in chapter 1, osmoregulators are animals which have the ability to maintain the osmotic pressure of their body fluids at levels which differ from those of the surrounding environment. The man-made distinction between osmoconformers and osmoregulators is of course an arbitrary one; there is a complete range of regulatory abilities ranging from "pure" osmoconformers, through animals having lesser or greater osmoregulatory capacities, to the brine-water osmoregulators described at the end of this chapter.

The shore crab, *Carcinus maenas*, is an example of the most common, and probably the most primitive, type of osmoregulation found in most osmoregulating invertebrates of marine origin. At high salinities the body fluids are iso-osmotic with the environment; the tissues must tolerate changes in extracellular fluid osmolarity and have the ability to change intracellular amino acid concentrations in the same way as the tissues of osmoconformers. The important difference is seen at low salinities; below a certain osmolarity the body fluids remain hyper-osmotic to the environment (figure 3.2). Problems of diffusional salt loss and osmotic water entry must thus be overcome. These problems can be alleviated by reducing the permeability of the integument to salts and/or water. Water permeabilities tend to be lower in brackish-water than in marine species, and some animals are able to reduce them still further as the external salinity decreases, but the necessity to expose a large surface area, relative to the body volume, to the environment for gaseous exchange means that the possibilities for reducing water exchange are somewhat limited.

Excess water must be eliminated as urine which, if iso-osmotic with the body fluids, will cause an additional loss of salts. Some species, e.g. the amphipod *Gammarus duebeni*, the shrimp *Crangon vulgaris* and the polychaete *Nereis diversicolor*, can produce hypo-osmotic urine, but most brackish water animals produce iso-osmotic urine. Urinary salt excretion is often only a small proportion of the total loss from the body; hypo-osmotic urine tends to be produced by animals with a reduced integumentary salt permeability and a high rate of water turnover. On the other hand, animals producing iso-osmotic urine often have low integumentary water permeabilities so that urinary flow rates are low and urinary salt losses are relatively insignificant. Low permeabilities to water and to salts do not necessarily go together.

Whichever methods of limiting water uptake and salt loss are in operation, there will always be some salt loss from animals osmoregulating in hypo-osmotic media, and it must be replaced by active uptake. Salt uptake sites are usually on the surface of the animal, either generally distributed as in polychaetes or localised upon the gills as in crustaceans and most teleost fish. Sodium uptake across *Carcinus maenas* gills is triggered when the haemolymph concentration falls below $400 \, \text{mmol} \, \text{l}^{-1}$. The rate of transport does not seem to vary down to the limiting external sodium concentration of about $70 \, \text{mmol} \, \text{l}^{-1}$ and the passive influx, which forms a significant proportion of the total influx, decreases with decreasing external salinity, so whilst *Carcinus* is able to maintain the haemolymph concentration above that of the external medium it is not able to keep it

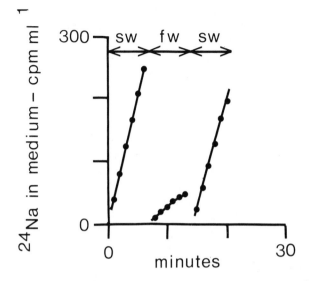

Figure 3.10 Outflux of radioactive sodium from a brine shrimp, *Artemia salina*, during transfer from sea water to fresh water and back (redrawn and modified from Thuet *et al.*, 1968).

constant (Shaw, 1961). However, in some species a high proportion of the salt uptake may occur across the gut wall, from either food or imbibed water. Salt uptake mechanisms will be discussed in more detail when freshwater animals are considered in chapter 4.

A number of brackish-water crustaceans show a more sophisticated type of osmoregulation in that they not only maintain hyperosmotic body fluids when in low salinities but can also keep them hypo-osmotic when exposed to high osmolarities. The most striking example is the brine shrimp *Artemia salina* (figure 3.10)—see section on hypersaline waters. Most euryhaline vertebrates osmoregulate in this way, including a number of fish species found in brackish water, but these will be discussed with the migratory species in chapter 5.

Urea retention

One further method of osmoregulation in brackish-water (and marine) animals is the use by some vertebrates of high body fluid urea levels to reduce osmotic stress. The elasmobranchs (the sharks and rays) originated in fresh water and have the reduced plasma salt concentrations characteristic of freshwater animals (see chapter 4). When they later colonised the seas, they solved their osmotic problems by retaining sufficient urea, $CO(NH_2)_2$, and trimethylamine oxide, $(CH_3)_3NO$, in their body fluids to make them very slightly hyperosmotic to sea water, thus allowing them to gain small amounts of water by osmosis (see chapter 5).

A few elasmobranch species live in brackish water or fresh water, including the sawfish *Pristis microdon*, which lives in the sea and in estuaries in south-east Asia and was investigated in 1930 by one of the founders of modern renal physiology, Homer W. Smith, who found that *Pristis* penetrated well into fresh water and under these conditions plasma urea concentrations were lowered and urine flow rate was very high—$250\ ml^{-1}\ kg^{-1}\ day^{-1}$. He concluded that *Pristis* was normally virtually iso-osmotic with sea water as a result of urea retention. As it moves into lower salinities urine flow increases to balance the osmotic water entry. Since most, but not all, the filtered urea is reabsorbed by the kidney tubules, renal urea loss becomes significant as urine flow increases and plasma urea levels fall thus reducing the animal's osmotic problems. As the fish swims back towards the sea, water entry and consequently urine flow rate decrease. Renal loss of urea diminishes and plasma levels rise until a point is reached at which the body fluids are again iso-osmotic with the environment. A very simple physiological mechanism thus accounts for the reduced body fluid urea levels in sawfish in fresh water which might otherwise have been mistakenly explained as the result of the reversal of an evolutionary trend.

Perhaps the most unexpected animal to be found in brackish water is the crab-eating frog, *Rana cancrivora*, which lives in mangrove swamps. As well as having low blood salt concentrations, most amphibia have very permeable skins and cannot survive in hyperosmotic media. *Rana cancrivora* maintains its body fluids slightly hyperosmotic to the environment by retaining urea, although it can also tolerate higher blood ion concentrations than can other amphibia. Adult frogs can survive in osmolarities as high as $800\ mOsm\ l^{-1}$ if acclimatised gradually and the tadpoles are even more tolerant—they need to be as they cannot hop out of the water and sit on the roots of a mangrove tree, as the adults do if the salinity gets too high.

Several other vertebrates have evolved urea retention mechanisms to enable them to live in brackish water. Amongst the reptiles the diamond-back terrapin, *Malaclemys centrata*, is confined to brackish water and has high plasma urea levels, unlike the freshwater members of the same family, the Emydidae.

Hypersaline and brine water areas

Relatively few species inhabit waters of more than $1200\,mOsm\,l^{-1}$. In terms of numbers of species per habitat, which presumably reflects the hostility of the environment, the following holds:

$$\frac{sea}{water} > \frac{fresh}{water} > \frac{brackish}{water} > \frac{hypersaline}{water} > \frac{brine}{water}$$

Invertebrates appear to resist high salinities better than do vertebrates. The teleost *Cyprinodon macularius* has been found in water of $2600\,mOsm\,l^{-1}$ and eggs of the related *C. variegatus* can hatch in $3200\,mOsm\,l^{-1}$ but the limit for fish appears to be about $2900\,mOsm\,l^{-1}$. Few invertebrates can survive above $5800\,mOsm\,l^{-1}$ but a handful of species, including the brine shrimp *Artemia salina* and the copepod *Tigriopus fulvus*, can, and do, live in saturated sodium chloride solution; in fact the former is common in brine pans used to produce salt by evaporation of sea water. One report describes *Artemia* living in a pool in the Crimea which contained $271\,g\,l^{-1}$ of salts; the water was said to have the colour and consistency of beer.

Many osmoconformers survive in up to $2000\,mOsm\,l^{-1}$, for example tropical bivalves living in salty lagoons. At higher salinities hypo-osmotic regulation becomes essential, not only to preserve the necessary chemical conditions within cells but also because as salinity increases viscosity increases and oxygen solubility decreases. If the same happened to the body fluids, problems would arise in supplying oxygen to the tissues, whereas if the animal osmoregulates, the oxygen content of the body fluids can be well above that of the external medium since it is the oxygen tension or partial pressure, not the oxygen concentration, which drives gaseous exchange. For example, an external pO_2 of $20\,kPa$ ($150\,mm\,Hg$) will cause net movement of oxygen into body fluids with a pO_2 of $17\,kPa$ ($128\,mm\,Hg$) even if the external medium is so hypersaline that it only contains $2\,ml\,O_2\,l^{-1}$ at a pO_2 of $20\,kPa$, whereas the body fluids contain $6\,ml\,O_2\,l^{-1}$ at a pO_2 of $17\,kPa$.

The osmoregulatory mechanisms of *Artemia* have attracted great interest, in spite of the practical problems of analysing body fluids from an animal only $10\,mm$ long, because it can regulate its internal osmolarity when living in a very wide range of external salinities (figure 3.2). The haemolymph sodium concentration is maintained within the range 145 to $190\,mmol\,l^{-1}$ when the external sodium concentration is varied from 240 to $1900\,mmol\,l^{-1}$. This regulation does not depend on a low permeability of the body surface to sodium ions, since 40% of the body sodium exchanges with the external medium per hour in sea water, and this figure rises to 80% per hour if the sea water is concentrated four-fold (Thuet *et al.*, 1968).

If internal sodium concentrations do not change, then sodium influx must equal sodium outflux. This could be achieved passively if the haemolymph of animals in sea water has an electrical potential of about +25 mV relative to the external medium (see section on electrochemical gradients in chapter 1); the measured potential is in fact very close to this value. If the animal is placed in fresh water (sodium concentration < 1 mmol l^{-1}) sodium *influx* will obviously be greatly reduced. To maintain sodium balance, *efflux* of sodium must also be drastically reduced, as in fact happens (figure 3.10). As the sodium concentration of the external medium is reduced, the positive electrical potential in the haemolymph decreases, and eventually becomes negative. This reduces the passive efflux of sodium and prevents excessive loss of this ion; it accounts for changes such as those shown in figure 3.10.

The reason for these changes in potential seems to be that the permeability of the gill epithelium to chloride ions is only about one-tenth of its permeability to sodium ions. If *Artemia* is placed in sea water, more sodium ions than chloride ions will diffuse into the body; the resulting positive internal potential limits further influx of sodium ions. In fresh water, sodium ions diffuse out down their concentration gradient faster than do chloride ions; the resulting negative internal potential limits sodium loss.

Because of this mechanism only chloride movements need be regulated actively; this requires a minimal energy expenditure because the chloride permeability is so much lower than the permeability to sodium ions. A large part of the chloride influx is coupled to its outflux by an exchange diffusion mechanism, but some active transport of chloride is necessary since the electrochemical gradient favours chloride entry in high-chloride media and its loss in low-chloride media.

Hypo-osmotic regulation leads to osmotic loss of water which has to be replaced by intake of sea water. *Artemia* kept in sea water drink a volume equal to 2 to 3% of their body weight per day. Absorption of this water from the intestine is a problem. As it is of higher osmotic pressure than the body fluids, water will tend to pass out of the animal into the gut lumen down the osmotic gradient, leaving the animal more dehydrated than if it had not drunk the water! In order to absorb sea water the dissolved salts must also be absorbed, adding to the animal's salt balance problems.

Artemia uses a mechanism similar to that evolved by marine teleosts which will be discussed in detail in chapter 5, and in essence involves active uptake of sodium chloride from the gut with water following passively. The sodium and chloride ions absorbed are then eliminated by the operation of the active chloride extrusion mechanism in the gills.

In hypo-osmotic media *Artemia* presumably eliminates excess water by urine production, but it is not a very convenient animal for studying this phenomenon. Its problems of water regulation are made easier by the fact that the permeability to water of its body surface is very much lower than that of other crustaceans—this must be part of the secret of its success.

CHAPTER FOUR

LIFE IN FRESH WATER

IN CONTRAST TO THE RELATIVELY CONSTANT MARINE ENVIRONMENT, conditions in fresh water can vary greatly and sometimes show marked variations over short time intervals. One reason for this is the small volume of most bodies of fresh water. Water in lakes, rivers, the soil and the atmosphere constitutes less than 0.025% of the earth's water. Of this amount a significant proportion is contained in a few large lakes, the North American Great Lakes complex and Lake Baikal in Siberia each containing about one fifth of the world's fresh water. Small volumes of shallow water often experience large changes in temperature and dissolved oxygen content in the course of a day. Salt concentrations are always very low compared to sea water but may show considerable variation from one body of water to another, although they are not usually subject to rapid fluctuations. In some circumstances, however, transient changes in ionic concentrations can be most pronounced. One example which has been studied is the way the hydrogen ion concentration can rise rapidly in the spring in Scandinavian rivers, often with disastrous consequences for the fish populations. This is due to pollution of the atmosphere with sulphur dioxide leading to an accumulation of sulphuric acid in the snow during the winter. In the spring the surface of the snow starts to melt and the water percolates down through the snow collecting sulphuric acid as it goes before running into streams which suddenly become very acid—as low as pH 4. This illustrates another contrast between sea water (which is well buffered and always slightly alkaline with a pH of around 8) and fresh water, where the buffering capacity depends on the nature and amount of the dissolved salts.

The salts in fresh water are derived mainly from atmospheric precipitation or by solution from rocks (other possible sources include wind-blown soil and pollution). Their concentrations may be affected by evaporation and precipitation. The osmolality varies from a few

$mOsm\ kg^{-1}$ downwards but is always much lower than that of the body fluids of any freshwater organisms, as are the concentrations of ions such as sodium and chloride (usually a fraction of a $mmol\ kg^{-1}$ but varying with distance from the sea; precipitation in coastal areas contains appreciable amounts of sodium chloride). The other main ionic constituents are calcium and magnesium ions (soft water from regions with igneous rocks contains very small amounts but they are the main cations in hard water from regions with calcareous rocks) carbonate and bicarbonate ions (derived from atmospheric carbon dioxide via precipitation or biological activity), sulphate and (in very small amounts) nitrate ions. "Average" ionic concentrations for world river water have been calculated as follows (Wetzel, 1975): Na^+, 274; K^+, 59; Ca^{2+}, 375; Mg^{2+}, 171; Cl^-, 220; $CO_3(+HCO_3^-)$, 479; NO_3^-, 17 $\mu mol\ l^{-1}$. However, levels can be as low as those found in some samples of water from the Amazon by Mangum et al. (1978): Na^+, 10; K^+, 20; Ca^{2+}, 3; Mg^{2+}, 2; Cl^-, 12 $\mu mol\ l^{-1}$. At the other extreme, concentrations up to and equalling those of brackish water (chapter 3) are found in many lakes.

Many osmoconformers have been able to colonise brackish-water estuaries but only osmoregulators have been able to move on into fresh water. No freshwater animals have body fluid osmolarities or salt concentrations which approach those of the environment. Consequently all face a severe problem in preventing dilution of the body fluids by both osmotic influx of water and diffusional loss of salts. The necessity of having a large surface area with very close contact between the body fluids and the water for oxygen uptake exacerbates the problems. In fact, if it were not for this requirement water entry would be of little importance, since non-respiratory areas of the integument may be almost completely imper-meable to water. Although oxygen is less soluble in sea water than in fresh water the upper layers of the sea are always saturated with air, and even at great depths the oxygen partial pressure is still high. Freshwater oxygen tensions may exceed that of the atmosphere due to photosynthetic activity during the day, but fall to low levels at night when utilisation exceeds supply. Increasing temperature decreases oxygen solubility and, in small volumes of water often leads to evaporation of significant amounts of water with consequent overcrowding of animals whose metabolic rates and oxygen requirements may have been considerably increased at the higher temperature. The respiratory organs of most freshwater animals must therefore be of sufficient surface area to allow oxygen uptake to proceed under unfavourable conditions, in spite of the osmoregulatory disadvantage of such an arrangement.

The following mechanisms are all involved in freshwater osmoregulation:

1. Reduction of the osmotic gradient between the body fluids and the environment.
2. Reduction of the permeability of the integument to water and salts.
3. Use of excretory organs to eliminate the excess water which enters by osmosis.
4. Active salt uptake mechanisms.

These will be considered in turn.

Reduction in body fluid osmolarity

No freshwater animals have retained the 1000 mOsm kg^{-1} body fluids of their marine ancestors—most maintain an osmolality of 300 or less, thus reducing the gradients for water entry and salt loss by at least two-thirds. As shown in chapter 1 this still enables them to maintain what appear to be optimum intracellular salt concentrations. But many animals function perfectly well with lower body fluid osmolalities. Table 4.1 gives a range of osmolalities and ionic concentrations which may be found in the extracellular fluids of freshwater animals. Simple animals, such as Protozoa with their large surface area to volume ratios, have internal osmolalities of around 100 mOsm kg^{-1} made up partly by organic molecules, but with considerably higher sodium and potassium concentrations than those of fresh water. More advanced animals, particularly those with well-developed neuromuscular systems, tend to have osmolalities in the 200–300 mOsm kg^{-1} range, although some invertebrate nerve cells are able to adapt to large changes in extracellular fluid ionic composition (Treherne, 1980).

Table 4.1 Examples chosen to illustrate the range of extracellular fluid osmolalities and ionic concentrations which may be found in freshwater animals. Freshwater mussels have the lowest values so far recorded, presumably a consequence of the very high surface area to volume ratio of their gills, and freshwater crabs the highest; they are thought to be relatively recent colonists of fresh water.

	mOsm kg^{-1}	ions (in mmol l^{-1} or kg^{-1})					
		Na$^+$	K$^+$	Ca^{2+}	Mg^{2+}	Cl$^-$	HCO$_3^-$
Margaritifera hembeli (mussel)	39	14.6	0.3	5.2	—	9.3	11.9
Lymnea trunculata (snail)	151	49	2.4	8.3	4.2	32.1	18.4
Lampetra fluviatilis (lamprey)	249	122	3.2	2.4	1.1	108	—
Aeshna granda (dragonfly larva)	395	145	9.0	7.5	7.5	110	15
Potamon niloticus (crab)	506	259	8.4	12.7	—	242	—

Reduction of permeability to water and salts

Water permeability

Freshwater animals tend to have lower permeabilities to water than do related forms living in brackish or sea water. Many freshwater crustaceans have permeabilities two orders of magnitude lower than those of their marine relatives. In teleosts, which also face osmotic problems when in sea water, permeabilities tend to be lower in marine animals. But comparisons between species are difficult, partly because of the technical difficulties involved in measuring water exchange. It is worth considering these in some detail in order to be able to make objective assessments of published data.

Water may be lost in urine or other secretions or in faeces and may be gained by drinking or in food (both directly and indirectly by metabolism) in addition to gain or loss across the body surface (dependent on the osmotic gradient). If the osmotic water flux cannot be measured directly (as is usually the case) it may have to be calculated by difference after all the other exchanges have been measured. Measurement of the rate of exchange of radioactively labelled water (tritiated water, HTO, which unfortunately diffuses slightly more slowly than H_2O across biological membranes; the difference is less than 10% and is usually ignored or forgotten) is a relatively simple procedure (described in chapter 11) but the values obtained often differ considerably from those obtained by calculating the balance of water gain and loss. Before considering the reasons for this discrepancy the question of how water moves across biological membranes and how this movement may be measured must be examined.

HTO flux is a measure of the diffusion of "labelled water" from one compartment (e.g. an animal's body fluids) into another compartment (e.g. the external medium). At the beginning of an experiment in which an animal has been injected with HTO the internal concentration of "labelled water" may be, say, $55\,mol\,kg^{-1}$ and the external concentration of "labelled water" zero. The magnitude of the initial rate of HTO outflux will therefore be related to this 55 molal concentration gradient. The measured volume flow of water between the animal and the environment will be related to the osmotic gradient, say $300\,mOsm\,kg^{-1}$ in a freshwater animal. The diffusional permeability coefficient, P_{diff}, can be calculated from the initial rate of HTO outflux thus:

$$P_{diff} (\text{in cm s}^{-1}) = \frac{F_{HTO}}{A[H_2O]}$$

$F_{HTO} = HTO$ outflux (mmol s^{-1}), A = surface area (cm^2) and [H$_2$O] = internal water concentration (mmol cm^{-3}). The osmotic permeability coefficient P_{osm}, can be calculated similarly:

$$P_{osm} \text{ (in cm s}^{-1}) = \frac{V}{A \sigma S}$$

V = net water flow (mmol s^{-1}), σ = reflection coefficient of solute and S = concentration gradient of solute (mmol cm^{-3}).

Estimating the surface area over which the exchanges take place is difficult, particularly in organisms with complex respiratory organs. The surface area of the gills of some very active fish, for example, is over 10 times the skin area. This adds to the problems of interspecies comparisons, but it might seem reasonable to suppose that in the same individual surface area is constant, so comparisons between P_{diff} and P_{osm} can be made without knowing the actual area. It is possible, however, that animals might be able to change the effective surface area of their gills by changing the patterns of blood and/or water flow to meet changing respiratory requirements.

The conflicting osmoregulatory and respiratory requirements of freshwater fish are well illustrated by the fact that increased oxygen consumption in trout is accompanied by increased urine flow, which presumably balances an increased osmotic entry of water. But in one set of experiments on freshwater rainbow trout, urine flow rate increased by only 22% when oxygen consumption doubled (Hofmann and Butler, 1979) suggesting that an increase in the effective surface area of the gills was not the major factor involved in the increased oxygen uptake. Hormones may be involved in these circulatory or ventilatory changes and they may also have a direct effect on membrane permeabilities. A valid comparison of diffusional and osmotic permeability coefficients can thus be made only if they are measured simultaneously or if they can be measured in an *in vivo* preparation of known area. Amphibian skins are very suitable for experiments *in vivo*; some results are shown in Table 4.2.

Table 4.2 P_{diff} and P_{osm} values for the skins of three amphibian species (from Maetz, 1968). Values in parentheses were obtained in the presence of neurohypophysial hormones.

	P_{diff} (cm sec^{-1})	P_{osm} (cm sec^{-1})	P_{osm}/P_{diff}
Rana esculenta	5.5×10^{-5} (5.7×10^{-5})	5.6×10^{-4} (1.43×10^{-3})	10.1 (25.1)
Bufo regularis	5.7×10^{-5} (5.9×10^{-5})	1.55×10^{-3} (1.95×10^{-3})	27.2 (32.8)
Xenopus laevis	3.0×10^{-5} (3.2×10^{-5})	2.8×10^{-4} (3.0×10^{-4})	9.3 (9.3)

Osmotic permeability often appears to considerably exceed diffusional permeability, and two explanations have been suggested.

1. The pore hypothesis Equality of the coefficients would imply that osmotic and diffusional water flow proceed by the same mechanism, i.e. random diffusion of water molecules. Net volume flow occurs in the presence of an osmotic gradient because diffusion from the region of high water activity (dilute solution) to that of lower water activity (more concentrated solution) exceeds diffusion in the reverse direction. It has been suggested that osmotic water flow proceeds not by random diffusion, but by bulk flow through aqueous pores in cell membranes. Consider the situation in figure 4.1A. A water molecule diffuses through a membrane by

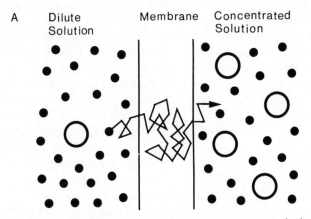

A   Dilute   Membrane   Concentrated
     Solution   Solution

Figure 4.1(A) Diffusion of a water molecule from a dilute to a concentrated solution. Small circles represent water molecules, large open circles solute molecules.

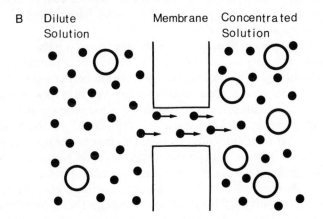

B   Dilute   Membrane   Concentrated
     Solution   Solution

Figure 4.1(B) Effect of a membrane pore on osmotic water movement—symbols as for figure 4.1(A).

random (Brownian) movement. More water molecules will diffuse from left to right than right to left, but the process will be slow, depending on the thermal energy of the water molecules. Figure 4.1B illustrates the possible consequences of the presence of a water-filled pore in the membrane. Water molecules leaving the pore to enter the solution are not all replaced by molecules from the solution. A pressure gradient is therefore set up, down which water molecules from within the pore pass. The motion of molecules along the pore becomes ordered instead of random, resulting in much faster flow. If flow is laminar, the pore resistance will be proportional to the fourth power of its radius (from the Poiseuille equation), so a small change in radius will have a large effect on flow.

If water molecules within pores are in a similar environment to those in bulk water, the activation energy for osmotic flow will be similar to that for self-diffusion in water, about 4.6 kcal mol^{-1}. This is the case with some cell membranes, e.g. those of dog red blood cells, and is good evidence that aqueous pores may exist in some cases. In an epithelium resembling frog skin which has been extensively studied—toad bladder—values of around 10 kcal mol^{-1} for the activation energy of HTO exchange have been obtained. Initially it was thought that neurohypophysial hormones reduced this to around 5 kcal mol^{-1} suggesting that they were opening up pores (their effect on the P_{osm}/P_{diff} ratio shown in some species, as illustrated in Table 4.2, could be explained by their increasing the radius of membrane pores) but later work showed that with efficient stirring this reduction in activation energy could be eliminated (Hays *et al.*, 1971). This illustrates the effect of unstirred layers; it has also been suggested that they could be the cause of the observed discrepancy between diffusional and osmotic permeability coefficients.

2. *Unstirred layers* Diffusion is a very slow process—the self diffusion coefficient of water is around 2×10^{-5} cm^2 sec^{-1} in the biological temperature range. The thickness of the diffusion barrier—body wall plus unstirred layers—has a great effect on water and also oxygen exchange. Most aquatic animals reduce this distance to a minimum by having a very thin respiratory epithelium, which is well stirred on both sides by circulatory and ventilatory mechanisms. This may not be necessary in very small animals with a high surface area to mass ratio, and different principles apply to the tracheal respiratory system of insects. The osmoregulatory implications of that system will be discussed when life on land is considered since although many insects are aquatic they have evolved from terrestrial ancestors.

Even very thin unstirred layers present a serious problem in studies of

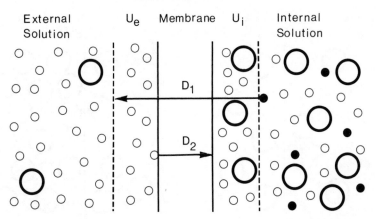

Figure 4.2 Effect of unstirred layers on movement of water across a membrane. Small closed circles represent labelled water molecules (HTO), small open circles unlabelled water molecules, large open circles solute molecules. U_e and U_i represent the external and internal unstirred layers. D_1 represents the effective diffusion distance for labelled, and D_2 the effective diffusion distance for unlabelled water molecules.

HTO fluxes. This is illustrated in a simple form, with a single membrane separating two aqueous compartments, in figure 4.2.

HTO added to the internal medium will mix rapidly if the bulk solution is well stirred. Sampling from a well-stirred external medium will enable the emergence of radioactivity from the external unstirred layer to be detected. Diffusion through the membrane plus unstirred layers will therefore be the rate-limiting step. The length of the diffusion path for HTO molecules is represented by the arrow D1. Both stirred and unstirred regions of the internal medium contain solute molecules, so osmosis will occur from the external to the internal unstirred layer. As soon as water molecules enter the internal compartment they will increase its volume, which is being measured to monitor osmotic water flow. The diffusion path for the water molecules involved in osmotic water flow is thus represented by D2. The longer the diffusion path through water the more the activation energy for HTO exchange will tend towards that for self-diffusion in water.

If water diffuses through the membrane at the same rate as through the unstirred layers, the calculated P_{osm}/P_{diff} ratio will depend on D1/D2. In fact, since the surfaces of most freshwater organisms are relatively impermeable to water, errors will arise only if the unstirred layers are large in relation to the membrane thicknesses. In respiratory organs such as fish gills, where the blood and water pathways are arranged to allow minimum

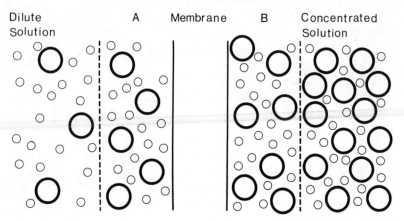

Figure 4.3 Effect of unstirred layers on the osmotic gradient across a membrane. Small circles represent water molecules, large circles solute molecules. A and B are the unstirred layers.

diffusion distances for oxygen, unstirred layers are of negligible importance. It has recently been realised that unstirred layers also affect osmotic flow, as illustrated in figure 4.3.

Water molecules leaving unstirred layer A will concentrate the solution within it, since the large solute molecules will be slow to diffuse back into the bulk solution. Similarly, water molecules entering unstirred layer B will dilute the solution within it. Concentration gradients will therefore exist across both unstirred layers, and the osmotic gradient across the membrane will be lower than would be expected from the composition of the bulk solutions.

The result of any comparison between P_{osm} and P_{diff} may well depend on which measurement is more erroneous! This makes it rather difficult to draw conclusions about the way water permeates membranes from physiological studies, and since any pores which exist are likely to be within membrane-spanning protein molecules, attempting to see them using electron microscopy is extremely difficult. Artificial lipid membranes, consisting of phospholipid bilayers similar to those found in cell membranes, do allow water molecules to pass through at a low rate. Two possibilities have been suggested; water molecules could either dissolve in the lipids and diffuse across, or pass through temporary dislocations in the liquid crystalline structure of the membrane interior produced by movements of the fatty acid chains of the phospholipid molecules. Such transient "pores", which would be very difficult to detect, might behave functionally in a similar manner to fixed pores, but this would make it difficult to explain the very high electrical resistances and very low salt permeabilities of artificial lipid membranes. However, whilst some biological membranes, e.g. those of freshwater protozoa or fish eggs, have water permeabilities as low as, or lower than, those of artificial lipid membranes, others are much more permeable, suggesting that aqueous pores may be present in some but not all animal membranes (Fettiplace and Haydon, 1980).

All the above considerations apply to a single membrane. The integument of an animal however often consists of a number of cell layers, with each cell offering several barriers to water movement—a cell membrane on either side, relatively unstirred cytoplasm in between, and sometimes extracellular layers such as basement membranes or surface mucous layers. The presence of unstirred layers is obviously an advantage from the point of view of osmoregulation, but a disadvantage from the point of view of respiration. However, recent evidence suggests that fluid movement across epithelia may take place mainly between the cells, rather than through them.

Salt permeability
Permeabilities to sodium and chloride ions, measured by flux experiments using ^{22}Na, ^{24}Na or ^{36}Cl, are much lower in freshwater than in marine animals. Large passive salt fluxes are no disadvantage to an osmo-conforming marine organism, and even when internal salt levels are maintained below those of the environment a high rate of exchange is not necessarily as disadvantageous as it may appear, as explained in the case of *Artemia salina* in chapter 3. But neither sodium nor chloride ions can be in equilibrium across the integument of a freshwater animal unless the transepithelial electrical potential is much higher than has ever been recorded. Reducing salt loss to a minimum limits the energy required for active uptake processes. It has been suggested that salt fluxes may be affected by respiratory requirements in the same way as water fluxes, but there would appear to be no reason why respiratory epithelia could not be almost impermeable to salts, as are artificial lipid membranes.

Some factors affecting permeability to water and ions
(*i*) *Temperature* A number of studies have shown Q_{10}'s (the relative increase of a parameter for a 10°C temperature rise) of between 2 and 4 for water fluxes.
(*ii*) *Environmental divalent ion concentrations* Calcium ions affect the

permeability and integrity of cell membranes and also the intercellular cements binding cells together. A number of studies have shown increased passive fluxes of both water and ions across fish gills if external calcium is removed; magnesium ions have a similar effect. An interesting example of the importance of the alkaline earth metals is given by experiments in which seawater-adapted eels were placed in artificial sea water lacking calcium and magnesium ions. They did not survive for more than five hours. After two hours the net osmotic flux had increased from -11.5 to $-39\,\mathrm{ml}\,\mathrm{h}^{-1}\,\mathrm{kg}^{-1}$, and there were also large increases in the passive fluxes of sodium and chloride ions. If however the eels were returned to normal sea water after three hours they recovered (Isaia and Masoni, 1976).

The calcium and magnesium concentrations of fresh water are lower than those of sea water and more variable, so it is important to remember that it is not only the osmolarity change which determines whether a fish transferred from sea water to fresh water will survive. Some otherwise stenohaline marine teleosts can live in fresh water if the calcium concentration is high enough to limit salt loss. Comparative data on permeabilities are thus useful only if the calcium and magnesium concentrations are known.

(*iii*) *Hormones*   Some animals are able to regulate the permeability of their body surfaces. Teleost fish do this by secreting the anterior pituitary hormone *prolactin*. In some species hypophysectomised animals can survive in iso-osmotic solution but not in fresh water, where they suffer a progressive and often rapid reduction in plasma osmolarity. Freshwater survival with maintenance of normal plasma osmolarity can be restored by prolactin injections. Normally prolactin is secreted in freshwater fish, but increasing the calcium concentration of the water to sea water levels can prevent this (Wendelaar, Bonga and van der Meij, 1980), and it seems that in this case the hormone is not required. *Adrenaline* increases branchial permeability to both water and oxygen in teleosts, as well as affecting gill blood circulation.

Water-excretory organs

There is a continuous osmotic inflow of water into all freshwater organisms and a variety of excretory organs have been adapted to remove it. Representatives of terrestrial air-breathing groups which subsequently adopted an aquatic existence, e.g. some insects, mammals etc., usually face much less severe problems because their body surface, which has no respiratory function, can be practically impermeable. But the first animals to enter fresh water must have done so from the sea via the brackish-water estuaries where, as we have seen, osmoconforming is an available option. In fresh water it is not, so presumably only those species possessing excretory organs which could be adapted for water elimination would have been able to penetrate into fresh water. One group is something of a mystery—the freshwater coelenterates osmoregulate but have no visible means of water excretion, although a hypo-osmotic fluid is excreted from the enteron (Prusch *et al.*, 1976).

In the supposed absence of active water transport, fluid elimination must depend either on solute secretion followed by passive water flow, or on filtration of body fluids—in fact ultrafiltration to retain valuable protein molecules. In either case the fluid initially produced contains salt concentrations similar to the extracellular fluid (differing slightly because of a Donnan distribution due to the absence of proteins). If this salt is lost from the body, as it is in a number of freshwater crustaceans which produce iso-osmotic urine, it has to be replaced by active uptake mechanisms at some other site. In most freshwater animals the bulk of the salts are reabsorbed before the fluid leaves the body. There are no obvious energetic reasons for adopting this mechanism. Although the work required to move an ion is proportional to the electrochemical gradient against which it must be transported, attempting to calculate the energy expended on ion uptake from consideration of the minimum work required involves gross oversimplifications. It leads to the conclusion that only a very small proportion of an animal's energy production is expended on osmoregulation, in contrast to the significant increases in metabolism observed in euryhaline animals when the osmotic gradients across the body wall are changed. The efficiency of the transport mechanisms will always be less than 100%. The sodium pump, for example, always seems to pump three Na^+ ions for each ATP molecule hydrolysed. The energy supplied is more than adequate to overcome the greatest adverse electro-chemical gradient encountered. If such a mechanism is involved in ion uptake, the energy expended in transporting one ion from a very dilute solution (e.g. fresh water) or a less dilute solution (e.g. urine) will be the same. Only the efficiencies of the two mechanisms will differ.

In reality the situation is much more complicated, as ions are absorbed into cells, not directly into the extracellular fluid. The electrochemical potential of the sodium ion within cells is usually very low, so uptake of sodium may not require any energy at all, being either by passive diffusion or by a Na^+/H^+ or Na^+/NH_4^+ exchange mechanism (see below). Work is done to maintain the low intracellular sodium concentration, not only in transporting epithelia but in all the tissues of the body. It is difficult to quantify the work involved, but according to one estimate (Keynes, 1975) roughly half the energy produced in the human body at rest goes to maintain the ionic gradients across the cell membranes.

Why then do most freshwater animals reabsorb most of the salts from the urine, in addition to having extrarenal uptake sites? As we will see when we consider active uptake mechanisms, the rate of ion transport is often limited by the very low ionic concentrations in fresh water. Presumably fewer pump sites are needed to maintain the same uptake rate from a more concentrated solution. We will consider three types of excretory mechanisms, illustrating different ways in which fluid is eliminated. Excretory organs which rely on salt transport to drive fluid secretion will be considered in chapters 5 and 8, when examples which have been extensively studied—aglomerular fish kidneys and insect Malpighian tubules—will be considered. For a wider range of examples see Riegel (1972).

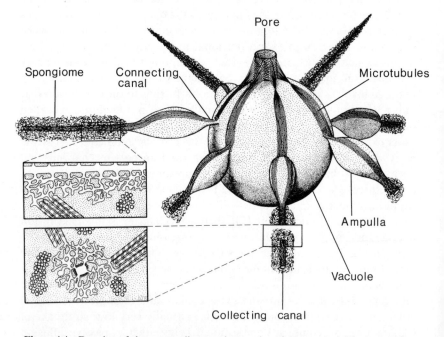

Figure 4.4   Drawing of the contractile vacuole complex of *Paramecium*. The inserts show longitudinal and transverse sections through the collecting canals to show the tubular system of the spongiome (drawn by D. J. Patterson).

(a) *Contractile vacuole complexes*

Contractile vacuoles are found in Protozoa and freshwater sponges. In some species they are transient structures which empty from time to time by fusing with the cell membrane. In others they are part of a more complex structure, including a permanent pore to the exterior through which fluid is expelled at intervals. In all cases there is a complex of tubules or vesicles, surrounded by many mitochondria, known as the spongiome, which surrounds the vacuole or its collecting canals (Patterson, 1980). The contractile vacuole complex of *Paramecium* is illustrated in figure 4.4.

The mechanism of fluid secretion is unknown, and likely to remain so for some time in view of the problems involved in studying such minute organelles, but where it has been possible to take samples of vacuole fluid by micropuncture it has proved to be of considerably lower osmotic concentration than the cytoplasm. The contractile vacuole complexes are thus able to excrete free water to compensate for osmotic entry. If the initial secretion is isotonic, salts must presumably subsequently be actively reabsorbed. Some marine protozoans have contractile vacuoles. Their function may be the elimination of water produced in metabolism, or which enters due to the colloid osmotic pressure of the cytoplasmic macromolecules or as a result of some ion uptake mechanism.

(b) *Vertebrate kidneys*

Some basic principles of renal physiology will be considered at this point, using the lamprey kidney as an example. Because a great deal of research has been directed at finding out how the human kidney works (or why it sometimes does not work), a vast amount of information is available about the mammalian kidney, although there are many aspects of it which are not yet understood. In contrast, we are in almost total ignorance about the mechanism of fluid formation in the contractile vacuole complex, and for all the varied excretory organs of the rest of the animal kingdom the state of our knowledge is intermediate between this and our knowledge of the mammalian kidney. Indeed, the physiology of invertebrate kidneys has often been deduced by analogy with the mammalian kidney. This has also often been true of lower-vertebrate kidneys. Although they have been studied for many years, it is only recently that some of the advanced techniques developed in medical research have been applied.

The evolutionary origin of the vertebrate kidney

There has been much controversy over a theory, one of whose proponents was the great pioneer of renal physiology, Homer Smith, that the

vertebrates must have arisen in fresh water because of the structure of their
kidneys. In his book *From Fish to Philosopher* he tried to illustrate the
evolution of the vertebrates, leading eventually to man, in relation to what
to him must have been the most significant organ in the body, the kidney!
It is an entertaining book but some of its arguments were a little far-
fetched. For example, he suggested that the heavy calcareous plates
covering the bodies of the early fishes were not armour to protect them
from predators but a means of isolating themselves from the osmotic
dangers of the freshwater environment. He forgot that these fish must also
have had gills in which the body fluids were brought into close proximity
with the environment for oxygen uptake. But the story is an interesting one
and, since the fossil evidence concerning the environment of the early
vertebrates may be equivocal (fossils found in marine deposits can easily be

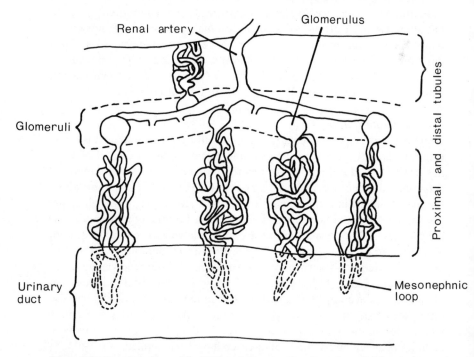

Figure 4.5 Drawing of part of the kidney of a lamprey, *Lampetra fluviatilis*, showing some of
the nephrons only for clarity. In reality the glomeruli are arranged in a continuous row along
the kidney. Only about 5 % of the length of the kidney is shown; on average 20 renal arteries
supply each kidney. Drawn from a "Microfil"-injected preparation. (*"Microfil" is a trade mark
of Canton Bio-medical Products, Boulder, Colorado.*)

explained away as having been washed down by rivers *post mortem*), it is worth re-examining the debate in the light of more recent comparative anatomical and physiological studies.

We obviously cannot study the ancestral vertebrates. The nearest available organism is probably *Branchiostoma* (formerly called *Amphioxus*). This has a system of nephridia which are unable to excrete the water which enters when the animal is placed in diluted (70%) sea water (Binyon, 1979). The most primitive vertebrates, the Agnatha, include the stenohaline marine hagfish and the euryhaline or freshwater lampreys. Both have glomerular kidneys derived from coelomoducts which are not homologous with the nephridia which their ancestors presumably possessed. The hagfish kidney, as described in chapter 2, is clearly primitive, having very few glomeruli arranged segmentally, although each has a much larger filtering surface area than other vertebrate glomeruli.

The situation is very different in lampreys. A 60 g lamprey has about 2000 glomeruli, which is comparable with the 2 000 000 glomeruli of a 60 kg human. They are arranged in a row down the centre of each kidney as illustrated in figure 4.5. All freshwater vertebrates have evolved this increased number of glomeruli. It seems that the possession of a glomerular kidney by their marine ancestors enabled the early vertebrates to colonise fresh water.

Glomerular filtration rate and clearance studies

To understand how kidneys work it is useful to be able to measure the glomerular filtration rate (GFR). The urinary papilla can be catheterised with polythene tubing and the rate of urine flow, $\dot{V}$, measured. This does not tell us the GFR, since fluid could be reabsorbed or secreted as the ultrafiltrate passes along the renal tubules. In the 1930s it was recognised by Homer Smith and others that the way to measure GFR was to identify a substance in the bloodstream (either endogenous or injected) which was completely filtered at the glomeruli but then neither reabsorbed or secreted by the tubule cells. GFR could then be calculated from the relationship

$$\mathrm{GFR} = \frac{U\dot{V}}{P}$$

where U is the concentration of the substance in the urine and P its plasma concentration. The most suitable marker substance then available was the plant sugar inulin, and this is still widely used, usually in a radioactively labelled form (either ^{3}H or ^{14}C), but other substances have also been advocated. Although early studies on aglomerular fish (see chapter 5) showed that inulin was not secreted by vertebrate kidney tubules, verification that it was not reabsorbed had until recently been carried out only on amphibian and mammalian kidneys.

The best way to discover whether or not a substance is absorbed is to inject a known quantity into a tubule (using a glass micropipette held in a micromanipulator) and see if it is all recovered in the urine. Tests on lamprey kidneys have shown that a small amount of the injected inulin is absorbed into the animal's blood stream (Moriarty et al., 1978). This means that inulin can be used only to give an approximate estimate of GFR (the error is about 16%) and illustrates the point (which has important consequences in osmoregulatory research) that a technique which is suitable for some animals is not necessarily applicable in others. In many instances, excretion of inulin has been taken as evidence for the occurrence of ultrafiltration, and in some cases the rates of appearance of inulin in the water following its injection into an animal have been used to calculate GFRs, in spite of conclusive evidence that inulin can pass across fish gills.

When a given volume of blood is filtered, all the inulin which it contains is removed, or cleared, from the blood. The GFR can be thought of as the volume of blood cleared of inulin per unit time, or the inulin clearance, C_{inulin}. The concept of clearance is a useful one in quantitative renal physiology. For any substance X,

$$C_X = \frac{[X]_{\text{urine}} \, \dot{V}}{[X]_{\text{plasma}}}$$

If any substance is completely cleared from all the blood plasma passing through the kidney its clearance will be a measure of the renal plasma flow (RPF). In mammalian studies p-amino hippuric acid (PAH) has been shown to fulfil this requirement since it is both filtered at the glomeruli and secreted by the tubules. Hence

$$C_{\text{PAH}} = \text{RPF}.$$

Unfortunately on the few occasions C_{PAH} has been measured in fish kidneys it has given such odd results as to be meaningless.

If a substance, for example glucose, is filtered but completely reabsorbed by the tubule cells, its clearance will be zero. The relative clearance of a substance, C_X/C_{inulin} is a useful concept. If the ratio is less than unity, the substance must have been reabsorbed; if greater than 1 the substance must have been secreted (assuming that it is filtered in the same way as inulin so that if it is neither secreted nor absorbed its relative clearance will be 1). The osmotic clearance, C_{osm} is defined thus:

$$C_{osm} = \frac{\text{urine osmolality} \cdot \dot{V}}{\text{plasma osmolality}}$$

A more useful concept, at least in studies on freshwater animals, is the free water clearance, C_{H_2O}:

$$C_{H_2O} = \dot{V} - C_{osm}$$

A positive free water clearance represents the volume of water (free from solutes) excreted by the animal. A negative C_{H_2O} represents the volume of water (free from solutes) gained by the animal as a result of the functioning of the kidney. Water gained in this way is of great significance in many birds and mammals, when water is in short supply.

The rate at which an individual glomerulus filters is the single nephron glomerular filtration rate (SNGFR). The nephron, the basic unit of the kidney, consists of the glomerulus, the Bowman's capsule and the kidney tubule, which can be divided into several regions (Logan *et al.*, 1980a)—see figure 4.6 which shows the parts of the lamprey nephron. It can be

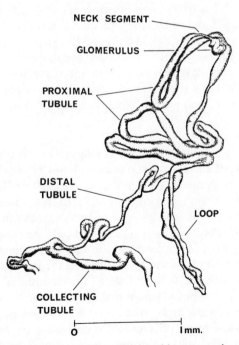

Figure 4.6 Drawing of a teased-out "Microfil"-injected lamprey nephron (from Moriarty, 1977, drawn by S. J. Moriarty).

measured in an animal previously injected with ^{3}H-inulin. An oil-filled micropipette is inserted into a tubule and a small droplet of oil injected, to be carried along the tubule by the flow of the tubular fluid. Suction is then applied to the pipette to keep the position of the oil drop constant and tubular fluid is collected over a timed period. SNGFR is calculated after assaying the radioactive counts, as follows:

$$\frac{\text{counts in sample}}{\text{counts per unit volume of plasma} \times \text{time}} = \frac{[\text{In}]_{\text{tubular fluid}} \cdot \dot{V}}{[\text{In}]_{\text{plasma}}}$$

Renal function in the river lamprey

Lampreys *Lampetra fluviatilis* make very good experimental animals because renal function is hardly affected by anaesthesia and the anatomy of the kidney makes micropuncture easy. The mean SNGFR is about 7 nl min^{-1} and the overall C_{inulin} (as we have seen a slight underestimate of the GFR) is about 60% of the body weight per day. The free water clearance is about half this figure because, although the urine osmolality (30 mOsm kg^{-1}) is very low compared to that of the plasma (250 mOsm kg^{-1}), almost half the fluid filtered is reabsorbed. The animals maintain a constant body weight although a volume of water equivalent to about 30% of the body weight passes in through the gills and out through the kidneys every day. An increase in the osmolarity of the external medium reduces the water influx and leads to a corresponding decrease in the mean SNGFR, the total GFR and the urine flow rate.

The lamprey nephron

The parts of the nephron are shown in figure 4.6; their contributions to the production of a hypo-osmotic urine will now be considered. The term glomerulus, strictly speaking, refers to a series of capillary loops which divide from the afferent arteriole and rejoin to form the efferent arteriole. They are almost surrounded by the Bowman's capsule, an extension of the renal tubule. Connective tissue (mesangial) cells are present between the capillaries. Figure 4.7 is an electron micrograph of a lamprey glomerulus showing the barriers between the plasma and the Bowman's space, which is continuous with the tubular lumen. The endothelial cells lining the capillaries are very thin and fenestrated, the large pores being no barrier to the movement of water and solutes. Next comes the basement membrane of the Bowman's capsule cells (or podocytes) and then the foot processes. Ultrafiltration takes place through the slit pores between these foot processes, probably along the whole length of the pores between the closely

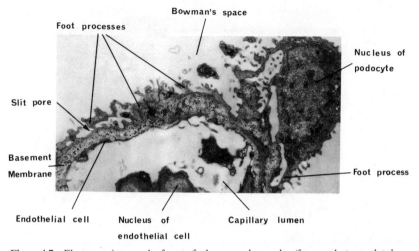

Figure 4.7 Electron micrograph of part of a lamprey glomerulus (from a photograph taken by A. G. Logan). Magnification × 5000.

applied glycocalyces of the podocytes. Filtration will occur if the hydrostatic pressure of the blood in the glomerular capillaries, about 2.7 kPa (20 mm Hg) in the lamprey, exceeds the colloid osmotic pressure of the plasma proteins (about 1.8 kPa) by an amount sufficient to force fluid through the filter. The filtrate then has to overcome the resistance to flow along the tubule but in lower vertebrates the cilia of the neck segment help to propel the fluid.

Studies by Richards in the 1930s on frog kidneys were the first to show that the fluid taken by micropuncture from the Bowman's space was an ultrafiltrate of plasma. Proteins are almost completely retained in the bloodstream but every compound with a molecular weight below about 40 000, including essential substances such as amino acids and sugars as well as salts, is filtered out. It has been said that the vertebrate kidney works on the principle of throwing the baby out with the bath water and then recovering the baby. Recovery of useful organic substances is carried out by the proximal tubule. In the lamprey proximal tubule little or no salt or water is reabsorbed—the tubular fluid over plasma inulin concentration ratio remains close to unity and the ionic concentrations, determined by electron probe X-ray microanalysis (see chapter 11), do not differ significantly from those of plasma. In the distal tubule most of the salts are reabsorbed (figure 4.8), together with some water. This may be an unavoidable consequence of the high osmotic gradient across the wall of

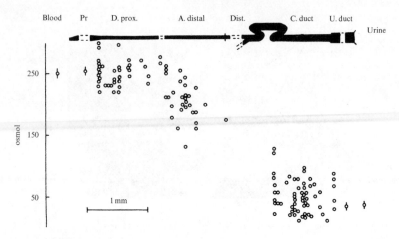

Figure 4.8 Variation in osmotic concentration (in mOsm l^{-1}) in tubular fluid as it passes down the tubule of the freshwater-adapted lamprey, *Lampetra fluviatilis*. Pr = proximal tubule (samples pooled, $n=18$, vertical bar represents $\pm$ standard error of mean). D. Prox. = descending proximal tubule, which forms first part of mesonephric loop. A. Distal = ascending distal tubule, which forms the latter part of the mesonephric loop. Dist. = remainder of distal tubule (inaccessible to micropuncture). C. Duct. = collecting duct. U. Duct. = urinary duct (mean $\pm$ S.E. given, $n=12$). Urine samples taken from catheterised urino-genital papilla (mean $\pm$ S.E. given, $n=9$) (modified from Logan *et al.*, 1980*b*).

the distal and collecting tubules. The end result is the elimination of large volumes of water with little loss of salts.

In teleost fish a urinary bladder is present and further reabsorption of salts takes place through its wall.

(c) *Crustacean antennal glands*

Crustaceans possess a pair of antennal glands, each of which has some resemblance to a vertebrate nephron. The anatomy is illustrated in figure 4.9. The coelomosac is lined by cells resembling those of the Bowman's capsule, complete with podocytes. It has an arterial blood supply, and filtration could occur as in the vertebrate kidney, driven by the hydrostatic pressure of the blood; Riegel (1977) has however proposed that an additional force is involved. He found what he called "formed bodies" in the lumen of the crayfish antennal gland and showed that they possessed the ability to attract water across a semi-permeable membrane. He suggested that they might be lysosomes secreted into the intercellular spaces between the podocytes. As the proteins they contained were

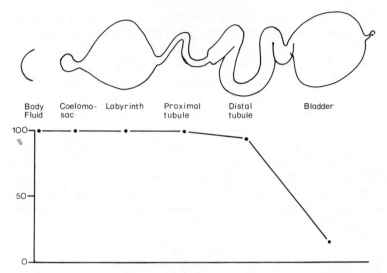

| Body
Fluid | Coelomo-
sac | Labyrinth | Proximal
tubule | Distal
tubule | Bladder |

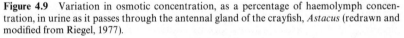

Figure 4.9 Variation in osmotic concentration, as a percentage of haemolymph concentration, in urine as it passes through the antennal gland of the crayfish, *Astacus* (redrawn and modified from Riegel, 1977).

degraded, their internal osmotic concentrations would increase above that of the extracellular fluid, causing entry of water. If the membranes of the "formed bodies" were impermeable to salts the salt concentration in the intercellular spaces would rise, causing osmotic movement of water across the basement membranes. The resulting hydrostatic pressure gradient would cause movement of fluid down the intercellular spaces into the coelomosac.

As in fish, salts are reabsorbed by the crayfish distal tubule and bladder cells resulting in very dilute urine (figure 4.9), although some crustaceans manage very well in fresh water without a distal tubule, producing iso-osmotic urine. Many marine fish also lack a distal tubule and cannot produce dilute urine.

Active uptake of salts

Sodium and chloride ions make up the bulk of the osmotic concentration of the extracellular fluid of all freshwater animals and, since neither ion is in passive equilibrium across the integument, diffusive and renal losses must be balanced by active uptake mechanisms. It is difficult to make direct measurements of the rate of ion uptake, so it usually has to be estimated by difference from studies on the balance of losses and gains. Renal losses can be quantified by collection of urine from cannulated animals. If a cannulated animal is placed in a known volume of fresh water the net extrarenal flux (gain − loss) of an ion can be followed by measuring the rate of change of its concentration in the external medium. Outflux from the animal can be measured by the use of a radioactive tracer (^{24}Na, ^{22}Na or ^{36}Cl for example) as described in chapter 11. Passive diffusion of ions from the very low concentrations found in fresh water into the animal will always be negligible, so net flux − outflux = active uptake. When active uptake is plotted against external concentration for sodium or chloride, graphs similar to that shown in figure 4.10 are found.

This resembles closely the relationship between the rate of an enzyme-mediated reaction and the substrate concentration which is described by the Michaelis–Menten equation:

$$V = \frac{V_{max} S}{K_m + S}$$

where V equals the velocity of the reaction at substrate concentration S and V_{max} and K_m are constants for any particular reaction. V_{max} is equal to

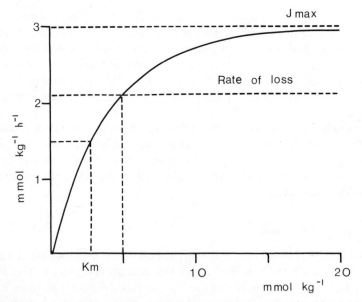

Figure 4.10 Relationship between rate of active ion uptake, in mmol kg^{-1} h^{-1}, and external ionic concentration, in mmol kg^{-1}, in a hypothetical freshwater animal.

the maximal rate of the reaction and K_m (the Michaelis constant, a measure of the affinity of the enzyme for its substrate) is equal to the substrate concentration at which the reaction proceeds at half its maximal velocity.

Presumably the kinetics of the ion-pump mechanism are similar to those of a substrate–enzyme mechanism, not surprisingly in view of what we now know about the sodium pump. The K_m is a useful criterion for characterising such a reaction and it is found that animals with low K_ms can live in more dilute solutions than those with higher K_ms, but the crucial factor for any animal is that the rate of uptake at the external concentration in which it is living should equal or exceed the rate of loss from the animal. In the example given in figure 4.10 the critical external concentration below which the animal cannot keep in positive sodium balance is 5 mmol kg^{-1}. (It must however always be remembered that salts ingested in food contribute to, and may be essential for, maintenance of ionic equilibrium.) Some examples of K_m and V_{max} (or J_{max} since a flow of any chemical is usually denoted by J) values are given in Table 4.3 (K_m and J_{max} can most easily be determined from a Lineweaver–Burk plot of $1/J$ against $1/$external concentration). The values given are a random selection of published data and give an idea of the efficiency of the uptake mechanisms of freshwater animals. Transport rates are obviously temperature-dependent; the measurements given were made at temperatures ranging from 10° to 24°C. In many cases the K_ms are similar to the normal environmental salt concentrations, or in other words the salt uptake mechanisms are operating on the rising part of the curve shown in figure 4.10, and the J_{max} cannot normally be achieved. On the other hand, following a short exposure to reduced salt concentrations the uptake rate will automatically increase after return to higher external concentrations, and any deficit which has been incurred will be corrected (Shaw, 1959).

Table 4.3 Values for K_m and J_{max} for sodium influx in a variety of freshwater animals

	K_m (mequiv/l)	J_{max} (mequiv/100 g/h)
Lumbricus terrestris (Annelida)	1.3	0.01
Lymnaea stagnalis (Mollusca)	0.25	0.0225
Aedes aegypti (Insecta)	0.55	1.2–1.3
Astacus pallipes (Crustacea)	0.25	0.095
Gammarus pulex (Crustacea)	0.1	0.22–0.31
Lampetra planeri (Cyclostomata)	0.26	0.036
Carassius auratus (Teleostei)	0.3	0.065
Salmo gairdneri (Teleostei)	0.02	0.065
Rana pipiens (Amphibia)	3–10	0.015
Xenopus laevis (Amphibia)	0.05	0.01

This contrasts with the situation in more "primitive" osmoregulators, which can tolerate brackish but not fresh water, such as *Carcinus maenas*, in which the active uptake rate remains at the V_{max} down to the lowest tolerable external concentration (chapter 3).

Sodium and chloride uptake mechanisms

Krogh showed that if fish or crustaceans were selectively depleted of either sodium or chloride (this can be achieved by keeping the animals in water containing salts with impermeant counter-ions such as sodium sulphate or choline chloride) and then returned to a solution containing sodium chloride, they showed an enhanced rate of uptake of only the ion in which they were deficient. He reasoned that since sodium and chloride uptake could proceed at quite different rates, each ion must be exchanged for another ion of the same charge to prevent a steady build-up of electric charge. Krogh proposed that two products, ammonium ions derived from protein catabolism and bicarbonate ions derived from carbon dioxide produced in respiration, were exchanged for sodium and chloride ions respectively, but demonstrating such exchanges has proved difficult. In fish, the gills are the site of nitrogenous excretion, ammonia being produced by deamination of amino acids in the gill cells as well as being carried to the gills in the blood stream and excreted there. One difficulty is determining whether it passes across the gills as free ammonia in solution or as ammonium ions. The reaction

$$NH_3 + H_2O \rightleftharpoons NH_4^+ + OH^-$$

is obviously affected by the pH of the solution, and the pH of the water immediately adjacent to the gills may differ from that of the bulk solution if CO_2 produced in respiration is forming carbonic acid. NH_3 may escape as gas from the water surface and NH_4^+ ions may be trapped in solution as NH_4HCO_3. In careful experiments it has been possible to demonstrate that in perfused trout heads ammonia can be excreted against a concentration gradient showing that something other than passive diffusion is involved (Payan, 1978). Injecting fish with ammonium salts stimulates sodium uptake, whilst adding ammonium salts to the external medium inhibits it and in some cases there is a good correlation between the rates of sodium uptake and ammonium loss.

All this indirect evidence supports the idea of sodium/ammonium exchange, but in many cases sodium uptake exceeds ammonium loss and it has been suggested that sodium ions may be exchanged for hydrogen ions. Sodium uptake and proton excretion can both be inhibited by the drug acetazolamide, an inhibitor of the enzyme carbonic anhydrase which catalyses the formation of carbonic acid from dissolved carbon dioxide. Dissociation of carbonic acid within the gill cells, which contain large amounts of carbonic anhydrase, may provide protons for exchange with sodium ions. In the goldfish there is an excellent correlation between the rate of excretion of ammonium ions plus protons and the rate of sodium uptake (figure 4.11), suggesting that in the absence of a sufficiently high rate of ammonia production, protons can be used to help balance sodium uptake. (Ionic regulation is obviously closely related to another vital function of fish gills, regulation of blood pH, but this topic is outside the scope of this book.) Amiloride, which blocks the entry of sodium into cells, has been shown to inhibit sodium uptake and ammonia excretion in the perfused trout head, suggesting that the exchange mechanism is located on the apical (outer) membrane of the gill cells.

Chloride uptake can be shown to be correlated with excretion of base (either bicarbonate or hydroxyl ions) and thiocyanate inhibits chloride uptake, in fact causing goldfish to lose chloride ions and gain carbon dioxide. In fish kept in sodium sulphate solution to prevent chloride uptake, carbon dioxide output is reduced and the carbon dioxide may even be gained from the water. This evidence suggests that the elimination of carbon dioxide, in the form of bicarbonate ions, may be obligatorily linked

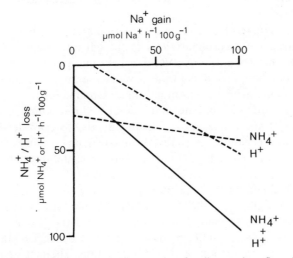

Figure 4.11 Relationship between active uptake of sodium and outflux of ammonium and/or hydrogen ions in the goldfish, *Carrasius auratus* (redrawn and modified from Maetz *et al.*, 1976).

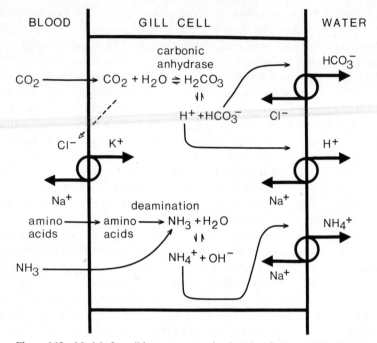

Figure 4.12   Model of possible transport mechanisms in a freshwater fish gill cell.

to chloride uptake. The conversion of respiratory carbon dioxide to carbonic acid and bicarbonate ions in the gill cells would provide the source of the bicarbonate ions and this is in keeping with the inhibition by acetazolamide of uptake of chloride as well as sodium ions. Some marine fish take up sodium ions in exchange for ammonium ions and chloride ions in exchange for bicarbonate ions despite the osmoregulatory disadvantages of these processes. Presumably their freshwater ancestors evolved these mechanisms to enable them to take up salts and they became dependent on them to get rid of nitrogenous waste and respiratory carbon dioxide.

Figure 4.12 summarises the processes thought to be involved in ion uptake by fish gill cells. The sodium and chloride ions presumably pass from the cells into the blood in the same way as they leave other cells; chloride ions by passive diffusion down their electrochemical gradient and sodium ions by the action of the sodium pump. The fact that ouabain inhibits sodium uptake and ammonia excretion in perfused trout heads is in accordance with this supposition.

CHAPTER FIVE

MOVEMENT BETWEEN FRESH WATER
AND SEA WATER

MOVEMENT OVER TWO VERY DISPARATE TIME SCALES WILL BE CONSIDERED: migrations during the course of evolution and during the lifetime of individuals.

A. Colonisation of different environments during the course of evolution

In the course of evolution many groups of animals have moved between marine, brackish-water, freshwater and terrestrial environments (see chapter 2). Sometimes the fossil record shows that these movements have occurred several times. Most freshwater animals are descended from marine ancestors via the intermediate habitat of brackish water.

Interpretations of evolutionary trends in osmotic adaptation may be difficult, but in one case it is possible to draw conclusions about the ancestry of some groups from a study of the osmoregulatory mechanisms of extant individuals. Evolution of reduced ionic concentrations in extra-cellular fluid is indicative of freshwater (or possibly terrestrial) ancestry. Present day marine animals showing this adaptation must have had a long period of evolution in fresh water, during which their bodies became so adapted to these low concentrations that they were unable to re-evolve the higher salt concentrations characteristic of indigenous marine animals. They had then to develop new mechanisms to overcome the osmotic and ionic problems of maintaining their plasma salt concentrations below those of the environment.

Two strategies have been employed. One is accumulation of organic solutes, mainly urea and trimethylamine oxide (TMAO), to keep the extra-cellular fluid iso-osmotic with sea water; this has been employed by the elasmobranchs, crossopterygians (as far as we know from the sole surviving representative, the coelocanth, *Latimeria chalumnae*), a few amphibia such as the crab-eating frog *Rana cancrivora*, and the brackish-water diamondback terrapin, *Malaclemys centrata*. The other is hypo-osmotic regulation—keeping the body fluids more dilute than the external medium and evolving mechanisms to cope with the resultant osmotic loss of water and diffusional entry of salts. This has been adopted by the marine lampreys and teleosts.

Air-breathing animals are isolated from their environments by an impermeable integument (see chapter 4). The problems encountered by, for example, aquatic birds or mammals when moving between fresh water and the sea are related to their dietary intakes of salts and water. Marine invertebrates always have extracellular fluid ionic concentrations similar to sea water. This applies even in the case of the truly euryhaline species *Eriocheir sinensis* which has reduced plasma ionic concentrations whilst in fresh water but allows them to rise when in the sea.

Iso-osmotic regulation

A "typical" marine elasmobranch is represented in figure 5.1; the plasma is in fact very slightly hyperosmotic to sea water, allowing a small osmotic influx of water to occur to replace that lost as urine and rectal gland secretion. Plasma sodium and chloride concentrations are greater than in freshwater fish but high levels of urea and TMAO account for much of the osmolarity.

In most aquatic animals nitrogenous waste is excreted as ammonia. This may be exchanged, in the form of NH_4^+ ions, for sodium ions in both freshwater and marine animals, but in any case its high solubility in water

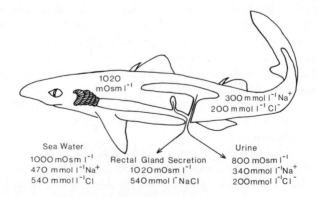

1020 mOsm l^{-1}

300 m mol l^{-1} Na$^+$
200 m mol l^{-1} Cl$^-$

Sea Water
1000 mOsm l^{-1}
470 mmol l^{-1} Na$^+$
540 mmol l^{-1} Cl

Rectal Gland Secretion
1020 mOsm l^{-1}
540 mmol l^- NaCl

Urine
800 mOsm l^{-1}
340 mmol l^{-1} Na$^+$
200 mmol l^{-1} Cl$^-$

Figure 5.1 Osmotic and ionic concentrations in plasma, urine and rectal gland secretion of a "typical" marine elasmobranch compared with those of sea water.

means that its disposal is no problem, in spite of its high toxicity. In elasmobranchs ammonia, derived from protein catabolism, is converted to urea, $CO(NH_2)_2$, by the enzymes of the ornithine cycle. Urea is less toxic than ammonia but concentrations well below those found in elasmobranch blood would prove fatal to most other animals. As well as evolving a tolerance to urea, elasmobranchs have developed a dependence on it, as illustrated by the fact that elasmobranch enzymes studied *in vitro* will function properly only in the presence of urea. Catastrophic effects of exposure of elasmobranch tissues to reduced urea concentrations include precipitation of eye lens protein and cessation of heartbeat.

Urea rapidly penetrates cell membranes, so exerts no osmotic effects on the cells, but losses from the body are minimised by low permeability of the gill cells and reabsorption by the kidney tubules. Few elasmobranchs live in fresh water. Homer Smith's studies on the sawfish, *Pristis microdon*, which is found in estuaries in south-east Asia and sometimes penetrates considerable distances up river, were mentioned in chapter 3. Since elasmobranch tissues are dependent on urea, life is very difficult for fresh-water elasmobranchs. Most of the species studied have quite high plasma urea concentrations $(100-200 \, \text{mmol} \, l^{-1})$ which increase the osmotic gradient across the gills and add to the osmotic problems. A few species of South American freshwater stingrays have re-evolved very low plasma urea levels and proteins which function normally under these conditions. TMAO contributes significantly to body fluid osmolarity in marine elasmobranchs but is probably derived from invertebrates in the diet.

Retention of urea and TMAO solves the osmotic problem of life in sea water but there is still the problem of ionic regulation, with sodium and chloride ions tending to diffuse in across the gills, and also being pumped in in exchange for ammonium and bicarbonate ions. Turnover of body sodium and chloride is very slow, less than 1% per hour, but a specialised sodium chloride excretory organ, the rectal gland, has evolved. Its secretion is iso-osmotic with plasma but contains very little urea or TMAO, in fact it consists almost entirely of a sodium chloride solution with considerably higher ionic concentration than plasma. Normally only 5 to 10% of sodium efflux is through the rectal gland, and studies showing that surgical removal had little effect cast doubts on its importance in osmoregulation. However, lip-sharks (*Hemiscyllium plagiosum*) force-fed on prawns showed a greater increase in total body sodium following rectal gland extirpation (Chan et al., 1967a), so it is possible that the gland plays a role in the elimination of excess sodium ingested in a diet of marine invertebrates. Rectal gland secretion is greatly stimulated by vasoactive intestinal peptide (VIP), thought to be released when food is ingested (Stoff *et al.*, 1979), so it seems likely that the function of the gland is to secrete excess salt entering through the gut. The gland contains one of the highest concentrations of $Na^+ + K^+$-activated ATPase found in any tissue; it is frequently used as the most convenient source of the enzyme for biochemical studies, but the mechanism of secretion is still poorly understood.

Hypo-osmotic regulation
The osmotic and ionic problems facing a "typical" teleost are illustrated in figure 5.2. In some, but not all, of the few species of marine teleosts so far

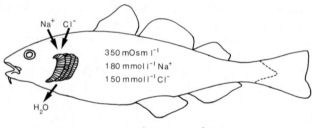

Na⁺ Cl⁻

350 mOsm l⁻¹
180 mmol l⁻¹ Na⁺
150 mmol l⁻¹ Cl⁻

H₂O

Sea Water   1000 mOsm l⁻¹ 470 mmol l⁻¹ Na⁺ 550 mmol l⁻¹ Cl⁻

Figure 5.2   Diagram to illustrate the osmotic and ionic problems faced by a "typical" marine teleost.

studied, the blood is about 20 mV (positive relative to sea water) and sodium ions are not far from equilibrium across the gills. The potential difference arises because the permeability to chloride ions is less than that to sodium ions, as described in chapter 3 for *Artemia*. Chloride ions will tend to move down their electrochemical gradient into the fish. Water will be lost osmotically across the gills and there will be a small renal loss. These losses can be replaced only by drinking sea water.

Water absorption by the gut
Marine teleosts drink continually, whereas freshwater teleosts drink little or no water. (Techniques for measuring drinking rates of aquatic animals are described in chapter 11). Drinking hyper-osmotic water does not immediately solve the fish's osmotic problems; it first adds to them. In seawater eels the oesophagus is highly permeable to sodium and chloride ions, but not to water, so salts diffuse into the body, reducing the osmotic concentration of the ingested water (Hirano and Mayer-Gostan, 1976).

The stomach receives partially "desalted" water, which is further diluted by diffusion of salts out of, and osmotic entry of water into, the stomach contents. By the time the fluid enters the intestine it has less than half the concentration of sea water, but is still hyperosmotic to the body fluids. The process thus far has resulted in a loss of water and gain of salts by the fish. There comes a point when loss of water into the intestinal lumen ceases and uptake into the body commences—apparently against an osmotic gradient. Skadhauge (1969) found that the more concentrated the medium to which an eel was adapted, the greater the ability of its intestine to absorb fluid against an osmotic gradient. Intestines from freshwater eels absorbed fluid if the adverse osmotic gradient was less than $73\,mOsm\,kg^{-1}$; those from seawater eels if it was less than $126\,mOsm\,kg^{-1}$, and intestines from eels adapted to an artificial double-strength sea water if it was less than $244\,mOsm\,kg^{-1}$. He found that the rate of fluid absorption was directly related to the net rate of salt uptake, suggesting that water was being transported as a direct consequence of active uptake of sodium chloride (Skadhauge, 1974).

Many instances have been documented of water moving across an epithelium in the absence of, or contrary to, an osmotic gradient. In virtually all these cases, water movement is dependent on active salt transport. Diamond suggested a mechanism for such "solute-linked water flow" based on the characteristic microanatomy of all such epithelia. He called his theory the "standing-gradient hypothesis" and proposed that water never in fact moved against an osmotic gradient, but that the establishment of local osmotic and hydrostatic gradients in lateral intercellular spaces resulted in net water movement (Diamond and Bossert, 1967). The model is shown diagrammatically in figure 5.3.

According to the theory, the sodium pump sites, instead of being uniformly situated around the periphery of the cell, are concentrated so as to pump sodium ions into the narrow parts of the intercellular spaces. The resulting high solute concentration in the spaces (high because the potassium ions exchanged for intracellular sodium ions diffuse back out again, whilst more sodium ions enter the cells from the luminal solution) produces a local osmotic gradient. Water flows down its concentration gradient from the bathing solution through the cells into the intercellular spaces. This will increase the hydrostatic pressure in the spaces, causing fluid to flow down the path of least resistance: along the gradually widening spaces and through the basement membrane, which poses no barrier to the movement of water and solutes. If the sodium pumps are confined to the narrow ends, and if the spaces are long enough, then osmotic

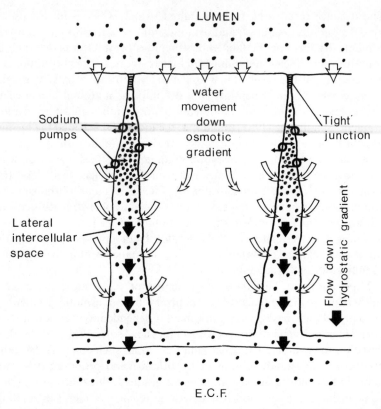

Figure 5.3 Diagram to illustrate the "standing gradient hypothesis" for solute-linked water flow across an epithelium. Density of dots proportional to ion concentration. Open arrows represent water movement, solid arrows movement of solution. ⚲ represents a sodium pump. E.C.F. = extracellular fluid.

equilibrium with the intracellular fluid will take place before the basement membrane is reached. It is proposed that this is the case in those tissues in which solute-linked water flow occurs in the absence of a macro-osmotic gradient across the epithelium (e.g. in gall bladder, which has been much studied). If equilibrium is not attained a hypertonic fluid will be transported, as is the case in fish intestine.

There have been a number of criticisms of this theory. It has been possible to demonstrate the hypertonicity of intercellular fluid in insect rectum, where the spaces are particularly large, both by micropuncture sampling of fluid and by electron probe X-ray microanalysis of deep-frozen hydrated sections. But if water is to enter the cells from the bathing solution the intracellular fluid should also be hypertonic to it. Recent evidence suggests that in many epithelia the so-called "tight junctions" are very leaky indeed, both to ions and larger molecules. They are in fact the main pathways for electrical conduction across "leaky" epithelia like those of kidney proximal tubules. The "standing gradient hypothesis" must therefore be modified to account for water movement through the "tight junctions". But how can osmosis occur through a pathway freely permeable to solute molecules? A number of theoretical studies of the model have either seemed to prove conclusively that it could not possibly work, or else proved conclusively that it could work perfectly well. Good evidence of the role of the intercellular spaces in solute-linked water flow comes from the elegant experiments of Spring and Hope (1978, 1979). *Necturus* gall bladder was mounted in a special chamber in which the intercellular spaces during fluid transport could be observed, and by focusing below the surface of the epithelium, the width of the spaces at different depths could be measured. The spaces were kept open by a very small hydrostatic pressure—about $3 \, cm \, H_2O$. Replacement of mucosal sodium led to the collapse of the spaces and shrinkage of the cells, due to the pumping out of all the intracellular sodium. Re-introduction of sodium to the bathing solution led to the return of the cell volume to normal (presumably as sodium entered the cells across their apical membranes) and reopening of the spaces. Resumption of sodium pumping into the spaces must have been followed by water movement and re-establishment of the hydrostatic pressure gradient.

However, as a recent review (Diamond, 1979) points out, "Although the evidence for an osmotic mechanism remains compelling, we don't know the relative importance of the transjunctional and transcellular routes of water flow, nor can we specify more than a lower bound on P_{osm}, nor do we know the forms of the osmotic gradients in the lateral intercellular spaces (if they exist), nor do we presently have methods capable of solving these problems. It seems futile to debate these questions further until someone thinks of new techniques to solve them."

Although the detailed mechanism of solute-linked water flow is still the subject of controversy it is clear that in many instances of water movement apparently against, or in the absence of, an osmotic gradient, including intestinal fluid absorption in marine teleosts, fluid flow is dependent on active solute transport. Fish are thus able to replace their water losses by drinking sea water, but this adds to their ionic problems as they then have to eliminate the ingested salts. The gills perform this function.

Salt secretion by gills

Keys demonstrated in the 1930s that the perfused head (including the gills) of the seawater eel secreted salts. In those days it was easy to measure chloride concentrations (by titration with silver nitrate solution), but there was no convenient analytical method for sodium ion, so everyone referred to chloride transport. Keys and Willmer (1932) suggested that cells found in marine teleost gills were the site of ion extrusion and called them "chloride cells". After the invention of the flame photometer, measurement of sodium concentrations became much easier, and everyone started talking about "sodium transport". A subsidiary reason was that two

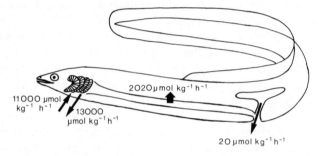

Figure 5.4   Sodium fluxes in an eel adapted to sea water (based on data from Maetz and Skadhauge, 1968, and Chester Jones *et al.*, 1969).

sodium isotopes, ^{22}Na and ^{24}Na, are relatively cheap and, being "hard" γ-emitters, are easy to radioassay; the only suitable radioactive chlorine isotope, ^{36}Cl, is expensive and more difficult to count, being a "soft" β-emitter, so relatively few chloride flux experiments have been performed. This explanation may seem less cynical when we consider how difficult it is to make the electrical measurements necessary to determine which ion is being actively transported.

The nature of the transport mechanisms in teleost gills is still the subject of much controversy, but it has become clear in recent years that the "chloride cells" (some would prefer to call them "ionocytes") are the site of sodium chloride extrusion, which balances the intestinal uptake and any net inward diffusion. But the net branchial excretion of sodium and chloride ions is lower by an order of magnitude than the outflux measured using isotopic tracers. Between 20 and 100% of the body sodium may be exchanged per hour, with influx being almost equal to outflux. Sodium fluxes in seawater eels are illustrated in figure 5.4.

Several explanations for these high fluxes have been advanced; they are not mutually exclusive and all may be involved to varying extents in different species.

1. With an internal electrical potential of $+20$ to $+25\,mV$, sodium ions may be in equilibrium across the gills, i.e. influx and outflux will balance, so large unidirectional fluxes are of no disadvantage to the fish. Chloride ions, in contrast, are far from equilibrium.

2. Influx and efflux may be linked by an "exchange diffusion" mechanism, possibly carrier-mediated. Such a mechanism may be part of an ion-pumping process. On theoretical grounds it may be deduced that a pump will operate more efficiently if it acts on the relative rates of influx and efflux of ion, rather than by a tight coupling between efflux alone and an energy source, since an irreversible reaction will have a much greater free energy change than a reversible reaction, and will require a correspondingly greater energy input to drive it (Fletcher, 1980).

3. Another explanation of exchange diffusion might be a pump which exchanges external potassium for internal sodium, but does not have an absolute specificity for external potassium. Although the affinity of the uptake process for potassium may be greater than that for sodium, because of the much higher concentration of sodium ($470\,mmol\,l^{-1}$) than potassium ($10\,mmol\,l^{-1}$) in sea water, it may pump in more sodium than potassium.

4. Sodium and chloride ions may pass freely through the junctions between neighbouring chloride cells (Degnan and Zadunaisky, 1980; Sargent et al., 1980; see figure 5.5).

Whatever the explanation of the high unidirectional fluxes, it is the net extrusion of sodium chloride which is important to the fish. Evidence that the active step is chloride transport comes from several sources. The most convincing experiments have been performed by taking advantage of the unusual location of some chloride cells in the killifish, *Fundulus heteroclitus*. In most teleosts chloride cells are found only in the gill filaments, but in *Fundulus* they are present in large numbers in the epithelium lining the operculum. This tissue can be dissected off and mounted between two chambers for measurement of isotopic fluxes and electrical parameters. The gills are far too complex for such experiments. In short-circuited (see chapter 11 for details of the method) *Fundulus* opercular epithelium, bathed on both sides with Ringer solution so that no electrochemical gradient exists for any ion, chloride is actively secreted and the net rate of chloride efflux exactly balances the short circuit current (Degnan et al., 1977). Chloride secretion and short circuit current are reduced by anoxia,

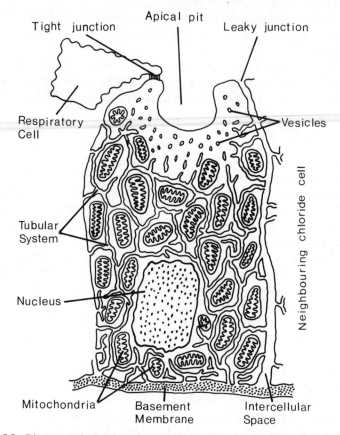

Figure 5.5 Diagrammatic drawing of a chloride cell from the gill of a marine teleost. (In reality there are many more mitochondria per cell and the tubular system is much more extensive.)

chloride removal, treatment with frusemide, the inhibitor of chloride transport, and also by ouabain, an inhibitor of $Na^+ + K^+$-ATPase.

$Na^+ + K^+$-ATPase is present in very large amounts in chloride cells, suggesting that the sodium pump is involved in salt extrusion. Chloride cells have a very characteristic ultrastructure (figure 5.5). Only a small surface area—the apical pit—is exposed to sea water. The cells contain many mitochondria, but the most typical feature is the extensively branching tubular system, which in effect is a vast extension of the basal and lateral membranes. Autoradiographic studies of ^3H-ouabain binding show that the $Na^+ + K^+$-ATPase is located in the membranes of these tubules.

Since ouabain binds from the extracellular fluid side of the cells, the pump sites would appear to be pumping sodium ions from the cells into the tubular lumens. It is not immediately obvious how this could contribute to the transport of chloride ions into the sea water. Several theories have been advanced, but data crucial to the understanding of chloride cell function is lacking, i.e. the cytoplasmic activities of sodium and chloride ions and the intracellular electrical potential. Measurement problems are associated with, firstly, the complex anatomy of the gills (although use of *Fundulus* opercular epithelium should obviate this problem) and secondly, the complex structure of the chloride cells. Electron probe X-ray micro-analysis has now reached the stage where intracellular ionic concentrations can be determined, but so far it lacks the resolution necessary to discriminate between the tubule lumens and the cytoplasm, and it would be difficult to determine whether microelectrodes inserted into chloride cells had entered tubules or cytoplasm.

In a number of epithelia, e.g. mammalian gall bladder and kidney tubules, chloride transport has been found to be dependent on sodium transport. In these two tissues intracellular chloride ion activities have been found to be greater than would be expected if the ion was distributed passively. It has been suggested that chloride enters cells against its electrochemical gradient by co-transport with sodium moving down its electrochemical gradient (Frizzell *et al.*, 1979), and it has also been proposed that this happens in chloride cells (Silva *et al.*, 1977). The sodium pumps would remove the sodium which entered the cells, and concentrate it in the tubules, maintaining a large gradient for passive sodium entry. Chloride ions entering with the sodium would not accumulate in the cell, because of the negative intracellular potential, but would pass out across the apical membranes (figure 5.6). This would explain why inhibition of

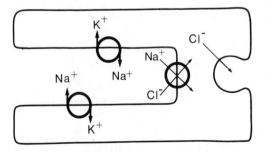

Figure 5.6 Diagram of a chloride cell to illustrate the model for chloride ion extrusion proposed by Silva *et al.*, 1977.

sodium transport with ouabain inhibits chloride extrusion. Research on chloride cells is at a fascinating stage at the moment and it must be emphasised that the mechanism outlined above is only one theory among many; it could be proved or disproved in the near future.

The role of the kidney
The kidney plays a very minor role in osmoregulation in marine teleosts and is often ignored. Urine flow rate is much lower than in freshwater fish and, since the need to produce large volumes of filtrate no longer exists, the number of glomeruli has often been reduced and in some species they have been lost altogether. One species with such an aglomerular kidney is the anglerfish, *Lophius piscatorius*, in which the whole of the renal arterial supply has also been lost and the blind-ending tubules are supplied with blood by branches of the renal portal vein. In perfused *Lophius* kidneys, urine flow rate and urinary magnesium concentrations in the perfusate rise with increasing magnesium concentrations in the perfusate, so magnesium secretion seems to be the driving force for fluid secretion (Babiker and Rankin, 1979). Figure 5.7 shows how magnesium excretion increases with increasing perfusate magnesium concentration. The function of the kidney

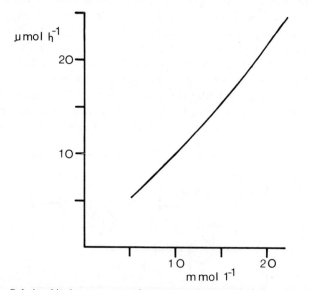

Figure 5.7 Relationship between rate of excretion of magnesium ions (μmol h^{-1}) and Ringer magnesium concentration (mmol l^{-1}) in an isolated perfused kidney preparation of the angler fish, *Lophius piscatorius* (based on data presented in Babiker and Rankin, 1979).

of marine teleosts seems to be excretion of excess divalent ions, which may possibly have been the function of the ancestral vertebrate kidney (see chapter 2). Urinary sodium and chloride concentrations are much lower than those of plasma, thus adding to the animal's salt load. (The fact that some sodium and chloride is excreted does not mean that the kidney is ridding the body of these ions; if the urinary concentrations are below those of the plasma, i.e. more water than salt is being excreted, the end result is an increase in the extracellular fluid sodium chloride concentration). The bladder continues the work of the kidney; sodium and chloride ions are reabsorbed, with water following, concentrating the residual magnesium and sulphate ions, while the sodium and chloride reabsorbed can easily be eliminated by the gills.

B. Migrations between the sea and fresh water

Many estuarine species exhibit varying degrees of euryhalinity, being able to cope with short-term salinity fluctuations as described in chapter 3. But very few migrate between the extremes of sea water and fresh water. Amongst the vertebrates which do are those fish which make anadromous (sea to fresh water) or catadromous (fresh to sea water) spawning migrations, such as salmon, some lampreys or eels. These few species have attracted much attention, and have taught us much about osmoregulatory mechanisms.

The eel will be used as an example; four closely related species have been much studied (the European eel *Anguilla anguilla*, the American eel *A. rostrata*, the Japanese eel *A. japonica* and the New Zealand eel *A. dieffenbachii*), which are very hardy and make excellent laboratory animals (the fishy equivalent of the white rat!). Eels breed in the sea and enter fresh water as elvers, which grow into yellow eels. After laying down large fat reserves these metamorphose into silver eels and migrate back to the sea to spawn and die. Both yellow and silver eels can withstand abrupt transfers between fresh water and sea water. We will consider the way the functions of the three osmoregulatory organs, the gills, gut and kidney, adapt to such changes.

The sodium/ammonium and chloride/bicarbonate exchanges of freshwater fish continue in sea water, so ion uptake following transfer to fresh water is probably no problem for any marine fish. Limiting ion loss is much more important. In an eel in sea water, sodium outflux, measured by the rate of appearance of radioactivity in the water from a fish injected with ^{24}Na, is very large. On transfer to fresh water there is an immediate

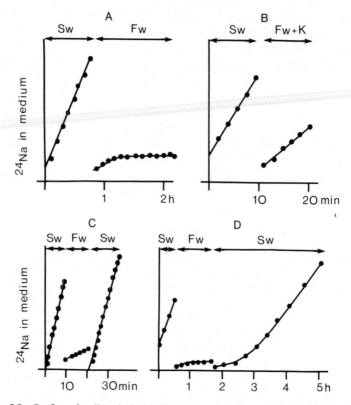

Figure 5.8 Outflux of radioactively labelled sodium ions from eels, *Anguilla anguilla L.*, subject to various changes of external medium. Sw = sea water, Fw = fresh water, Fw + K = fresh water + 5 mmol l^{-1} KCl (redrawn and modified from Motais, 1967, and Maetz, 1971).

reduction ("instantaneous regulation") followed half an hour later by a further reduction to a very low level ("secondary regulation") as shown in figure 5.8A. Stenohaline marine fish, which do not show these reductions, rapidly lose a large proportion of their body sodium and die. The instantaneous regulation is much less marked in fish transferred to fresh water containing low concentrations of potassium ions (figure 5.8B). The reduction in sodium outflux can partly be explained by changes in the transbranchial electrical potential, arising from the differential permeability to sodium, potassium and chloride ions, but suggests that sodium/potassium exchange may be involved in marine osmoregulation. Exchange diffusion, or a pump-leak mechanism, has also been implicated.

The secondary regulation is presumably the result of a permeability change, possibly under hormonal control. The anterior pituitary hormone prolactin is necessary to limit sodium loss in freshwater teleosts (see chapter 4).

If an eel is rapidly transferred from sea water to fresh water and then back again, the sodium efflux immediately returns to its seawater rate (figure 5.8C). Following a long period in fresh water, transfer to sea water only results in a gradual increase in sodium outflow (figure 5.8D). The number of chloride cells, and the branchial $Na^+ + K^+$-activated ATPase activity, also rises over a period following transfer. All these changes are dependent on the secretion of the hormone cortisol by the interrenal gland; figure 5.9 shows the changes in plasma cortisol concentration, gill ATPase activity and sodium efflux in *A. rostrata* (Forrest *et al.*, 1973). Caution must be exercised in interpreting any variations in plasma cortisol concentrations in fish, since they are elevated for several days following any stress, such as handling. However, it is significant that adrenalectomised European eels do not survive transfer to sea water unless given injections of cortisol (Mayer *et al.*, 1967).

Hormonal influences on the alimentary canal are also important during changes of environment. In the seawater eel the oesophagus is highly permeable to sodium and chloride ions, which allows the "desalting" of the ingested water, whilst in freshwater eels it is impermeable. Cortisol increases the permeability and prolactin decreases it (Hirano, 1980). Fluid absorption by the intestine *in vitro* increases following transfer of an eel from fresh water to sea water. Cortisol stimulates both fluid uptake and $Na^+ + K^+$-ATPase levels, but the enzyme is much slower to respond than the transport mechanism. A similar lack of correlation between changes in the enzyme activity and the uptake rate occurs after acclimatisation to different temperatures, so the picture is somewhat confusing. The river lamprey, *Lampetra fluviatilis*, rapidly loses the ability to osmoregulate in sea water once it has entered fresh water on its anadromous spawning migration, probably because the intestine starts to degenerate following cessation of feeding on leaving the sea.

As we have seen, freshwater teleosts produce large volumes of dilute urine which would obviously be quite inappropriate in sea water. Transfer of a rainbow trout from fresh water to sea water results in a rapid and dramatic fall in both GFR and urine flow (figure 5.10). Changes in GFR in teleosts seem to result from changes in the number of functional glomeruli, a process known as "glomerular recruitment". There is some evidence that the neurohypophysial hormones, arginine vasotocin and

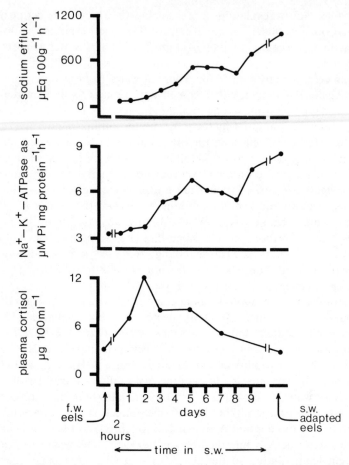

Figure 5.9 Changes in sodium efflux, gill Na⁺ + K⁺-activated ATPase activity and plasma cortisol levels in freshwater eels transferred to sea water (redrawn and modified from Forrest *et al.*, 1973).

isotocin, may be responsible for the reduced urine flow in seawater eels; injection or infusion of small amounts of either hormone produces antidiuresis (reduction in urine flow) (Babiker and Rankin, 1978). Angiotensin II (see chapter 9) has also been proposed as the factor regulating GFR in euryhaline teleosts, since it produces a glomerular antidiuresis in the trout (Henderson and Brown, 1980). Magnesium secretion in marine fish kidneys is stimulated by increased magnesium concentrations in the renal circulation, as described above.

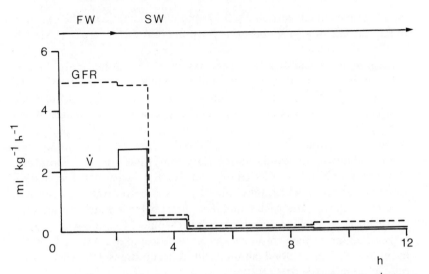

Figure 5.10 Changes in glomerular filtration rate (GFR) and urine flow rate ($\dot{V}$) in a rainbow trout, *Salmo gairdneri*, transferred from fresh water to sea water (Sinnot and Rankin, unpublished observations).

Few invertebrates can migrate between sea water and fresh water, and only one has been studied in detail; the Chinese mitten crab, *Eriocheir sinensis*. Adult *Eriocheir* spend most of their adult existence in fresh water, which may be moderately soft, though not as low in ionic concentration as that inhabited by crayfish such as *Astacus*. Because its larvae cannot survive low salinities, the mitten crab has to return to the sea several times in its life cycle to breed. *Eriocheir* has very low permeabilities to salts and water. Its water permeability is less than 10% of that of the euryhaline shore crab *Carcinus maenas*, which can tolerate water of about $200\,\mathrm{mOsm\,l^{-1}}$ indefinitely, and the mitten crab also has a lower permeability to water than either the eel (*Anguilla*) or the crayfish (*Astacus*). The very low permeability of *Astacus* to sodium however probably explains why it can tolerate softer water than can *Eriocheir*. In most respects therefore, the mitten crab can be thought of as a super *Carcinus*. Like the shore crab, *Eriocheir* is virtually iso-osmotic with full sea water (haemolymph osmolality $= 1100\,\mathrm{mOsm\,kg^{-1}}$), and starts osmoregulating in dilute media. In fresh water it has a blood concentration of $550\,\mathrm{mOsm\,l^{-1}}$, maintained by active salt uptake by the posterior gills. The low permeability of the integument allows *Eriocheir* to produce urine iso-osmotic with the haemolymph without requiring a vast urine volume (with consequent catastrophic salt loss) to offset osmotic uptake of water. In fact the filtration rate and urine flow rate of *Eriocheir* are both similar to those in *Carcinus*.

Because the haemolymph concentration of the mitten crab changes considerably during its migration (unlike the blood concentrations of the migratory teleosts) its tissues require efficient intracellular iso-osmotic volume regulation to prevent or reverse swelling or shrinkage. When moving from fresh water to sea water, the free amino acid content of the tissues roughly doubles. In nature this doubling probably prevents cellular volume changes, since migrations take some time. In the laboratory, however, an abrupt transfer from fresh water to sea water does induce tissue dehydration, which takes about two weeks to reverse. In contrast a transfer of seawater-acclimated *Eriocheir* to fresh water causes only transient tissue swelling, which is reversed within 24 hours. Amino acids rather than intracellular proteins appear to be lost to or taken up from the haemolymph, since blood amino acid concentrations fall when intracellular amino acid concentrations rise, and vice versa.

CHAPTER SIX

EMERGENCE FROM WATER

MANY AQUATIC ANIMALS ARE ABLE TO LEAVE THE WATER FOR LIMITED periods of time, for example littoral animals at low tide or animals in tropical ponds which dry out between rainy seasons. They face many problems, including loss of physical support from the surrounding aqueous medium, and exposure to freezing or over-heating; and their respiratory structures may become ineffective (gills tend to collapse in air). Many of these problems and their solutions are outside the scope of this text but perhaps the most serious one, the potential for desiccation with consequent concentration of the body fluids, is an osmotic one. Once these initial problems were solved the way was open for more completely terrestrial animals to evolve. Longer-term problems of absence from water included, apart from such non-osmotic problems as the evolution of new reproductive strategies, the elimination of nitrogenous wastes and the acquisition of salts.

Prevention of desiccation

The rate of evaporation from the body surface varies with a number of factors—the permeability of the integument to water, the establishment of local humidity gradients (this factor being strongly influenced by wind conditions), and the surface temperature (which will itself be influenced by the rate of evaporative cooling) but the humidity of the atmosphere is the most important. Dissolved solutes lower the vapour pressure of a solution. The degree of lowering is proportional to the reduction in the chemical potential of the water or, as expressed by Raoult's Law, to the mole fraction of the solute. The vapour pressure of pure water at 18°C is about 2.1 kPa (15.5 mm Hg). The presence of solutes in the body fluids of an iso-osmotic marine animal will reduce this by about 0.04 kPa (0.4 mm Hg). The animal will therefore lose water to air which is anything less than 98 %

saturated with water. Even a saturated solution of sodium chloride ($317\,\mathrm{g\,l^{-1}}$ at 20°) is in equilibrium with air only 75% saturated with water vapour, so a marine animal moving into dry air would, if it had no special modifications to limit water loss, be faced with a problem of desiccation almost 2 orders of magnitude more severe than it would face if immersed in a medium twice as concentrated as sea water. This illustrates the immense osmoregulatory problems involved in leaving the aquatic environment.

Most littoral animals and terrestrial animals of littoral origin are quite incapable of withstanding desiccation and simply avoid exposure to it. Littoral bivalves, barnacles and gastropods (e.g. winkles, limpets) preserve their aquatic environment by enclosing bodies of water within relatively impermeable structures which surround their vulnerable tissues. In analogous fashion, burrowing bivalves, crustacea and polychaetes (e.g. *Arenicola marina*), retreat into their burrows where they remain immersed in water, or at least in water-saturated air, until the tide returns to cover the substrate. On the lower portion of rocky shores, where tidal exposure is of shorter duration, rather less efficient strategies may be observed. Sponges and bryozoa live amongst damp weed, while sea anemones inhabiting damp crevices and overhangs contract to reduce their surface area and rate of water loss. Some shore fish, such as the shanny, *Blennius pholis*, also live in damp crevices, relying upon inactivity and cutaneous respiration to survive in circumstances where the gills are ineffective.

Nitrogenous excretion

Elimination of nitrogenous waste, derived mainly from protein catabolism, is a major problem associated with emergence from water. Aquatic animals have no problems, since the highly toxic ammonia produced can rapidly be eliminated across the gills. Its conversion to less toxic compounds requires the expenditure of energy, so even terrestrial animals excrete ammonia under conditions where it can quickly be eliminated from the body. The distal tubule of the mammalian kidney, for example, is the site of ammonia production and excretion since urine passes quickly from thence to the bladder, the walls of which are relatively impermeable.

Terrestrial members of the Amphibia provide a good example of the relationship of nitrogenous excretion to water availability. The aquatic tadpoles have no problem in excreting ammonia across the gills, but during metamorphosis the enzymes of the ornithine cycle are induced in the liver by the hormone thyroxine, where they convert ammonia to urea.

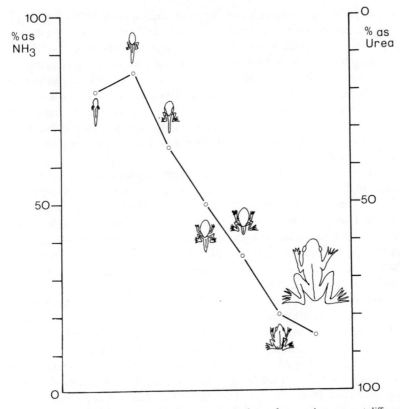

Figure 6.1 Percentage of nitrogenous excretion in the form of ammonia or urea at different stages in the metamorphosis of tadpoles of *Bufo bufo* into toads (based on data in Munro, 1953).

Figure 6.1 illustrates the changing pattern of nitrogenous excretion during amphibian metamorphosis. The urea formed is excreted renally, but at the expense of an appreciable loss of water (since amphibians cannot produce urine of greater osmotic concentration than their body fluids) and other solutes apart from urea will be present.

The emergence of terrestrial animals

A. From sea water

Because the majority of terrestrial animals appear to have arisen from forms which have followed the sea water → brackish water → fresh water route before invading the land, the more direct route of invasion, across the littoral zone, has attracted rather less attention. Although it is generally true that terrestrial animals which have a littoral rather than a freshwater ancestry (e.g. isopods, amphipods, land crabs, some gastropods) are less well adapted to a terrestrial existence in that they have a greater dependence upon damp surroundings, they nevertheless make a substantial contribution to the fauna in some areas. It must also be remembered that our knowledge of the evolution of some very successful terrestrial groups (e.g. nematodes, oligochaetes, arachnids, gastropods) is insufficient at present to tell us with certainty whether they reached land from the sea or via fresh water.

In response to fluctuating salinities in the littoral zone (see chapter 3), many of its inhabitants have become efficient osmoregulators, especially the crustaceans (e.g. crabs, isopods and amphipods) and gastropods. In combination with behavioural responses to humidity and light, which keep them in damp air beneath shelter from the influence of sun and wind (which potentiate desiccation to a considerable extent), this osmoregulatory ability provides powerful pre-adaptation for a more permanent terrestrial existence. Thus many terrestrial species such as woodlice (isopods) and rain-forest amphipods differ very little in their form and physiology from their littoral relatives living beneath pebbles and seaweed. Many animals inhabit microenvironments where the humidity is at or approaching saturation, e.g. beneath stone or leaf litter or in burrows in the soil (although water can be lost to the soil if the water potential of the soil is less than that of the body fluids), or even lead an aquatic existence, e.g. in surface films of water. The simplest and most widespread strategy for avoiding desiccation is the adoption of nocturnal behaviour. As the ground cools at night, the humidity of the layer of air adjacent to it increases, often becoming supersaturated with dew formation. Other behavioural adaptations include the ability, shown for example by woodlice, to detect and move along humidity gradients. The only physiological adaptation that separates many of these animals from their more aquatic relatives is a tendency to excrete nitrogenous waste as urea or uric acid rather than as ammonia.

The semi-terrestrial and terrestrial land crabs provide us with examples of greater resistance to desiccation. Firstly, these crabs are characterised by reduced cuticular permeability to water. Thus, while the aquatic blue crab *Callinectes sapidus* loses about $15 \, mg \, H_2O \, cm^{-2} \, h^{-1}$ through its carapace, the land crab *Gecarcinus lateralis* loses only $1.2 \, mg \, H_2O \, cm^{-2} \, h^{-1}$ under the same experimental conditions. Interestingly, virtually all water loss from land crabs is from the legs and body; water loss from the gill chambers is negligible. This reduced permeability has been understressed in the past, probably because the permeability is still very high in comparison with terrestrial arthropods; land crabs lose water at 6–7 times the rate of cockroaches. However, to put these rates into perspective, the highly adapted desert kangaroo rat loses water at about 25% of the land crab rate.

Like terrestrial isopods and amphipods, the land crabs exhibit behavioural mechanisms to conserve or obtain water, most spending considerable periods in damp burrows and also visiting pools regularly to drink. Some species have the ability to discriminate between low- and high-salinity water and tend to drink the former. Many can also pick up water from a damp substrate by capillary action along tufts of hairs, which lead to the branchial chamber where water is absorbed across the gills. Finally, several of these crabs have enlarged pericardial sacs which store water. This storage is not primarily for routine water conservation, but for providing fluid for expansion of the new exoskeleton after moulting.

B. From fresh water

Many terrestrial animals are descended from freshwater rather than marine forms and there are several reasons why this should be so. As mentioned in chapter 4, oxygen shortage is much more likely in fresh water than in the sea, and in the course of evolution many freshwater organisms have taken advantage of the plentiful supply of oxygen available close at hand in the atmosphere, whilst still retaining their aquatic existence. Many interesting examples of this phenomenon can be found in the Amazon basin. During the rainy season, the river floods vast areas which then slowly dry out exposing many aquatic organisms to overcrowding and, during daytime, to severe oxygen depletion as oxygen levels fall and metabolic rates rise as high temperatures are reached in small bodies of water. A great variety of air-breathing species has evolved in these circumstances. A number of different organs have evolved for aerial respiration and with their advent one of the great problems to be overcome in emerging from water was solved. An unavoidable consequence of air-breathing, however, is the loss of large volumes of water by evaporation from the moist respiratory surfaces (although for an animal living in fresh water, replacement of the lost water would be no problem). An additional risk of desiccation that would never be encountered by marine species is that small bodies of fresh water may dry up altogether, for prolonged periods at times particularly in the tropics. Survival under these conditions depends on limiting water loss and tolerating partial dehydration.

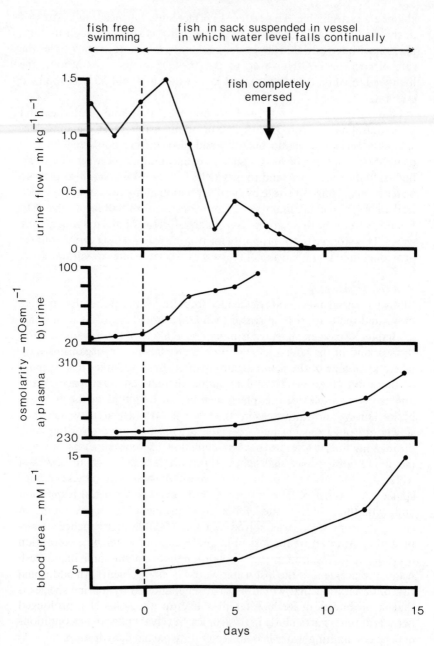

Aestivation in lungfish

A good example of the way an animal, related to the ancestors of terrestrial vertebrates, copes with these problems is the African lungfish, *Protopterus*. As water evaporates in the dry season, lungfish become trapped in shallow pools. As soon as they are exposed to air they burrow into the mud, going deeper as the water level in the mud falls, and leaving a narrow air passage to the surface. They then fold themselves up with their tail over their head and secrete a mucus cocoon around themselves, leaving a small hole above the mouth to breathe through, and go into a state of dormancy known as aestivation. Some of the changes occurring during the first few days of aestivation are shown in figure 6.2 (in this example artificially induced in the laboratory). Urine production declines by more than 50% as blood pressure falls and eventually urine flow ceases. Lungfish remain completely anuric for long periods, up to several years, of aestivation. In water they lose 65% of their nitrogenous waste as ammonia across the gills, the remainder being converted to urea before being excreted. During aestivation only urea is produced and since it cannot be excreted by either the gills or the kidneys it accumulates in the body—*Protopterus* is yet another animal whose tissues have evolved a tolerance to high urea levels. Urea may make up as much as 3% of the body weight during prolonged aestivation. Protein is the major energy source although metabolism is greatly reduced during aestivation, and as well as the nitrogenous waste, protein catabolism liberates sulphur (as sulphate ions) which accumulate in the extracellular fluid. Although enclosed in a waterproof cocoon the lungfish needs to breathe, albeit at a greatly reduced rate, so gradual evaporation of water from the respiratory surfaces leads to an increase in concentration of all ions in the body fluids. In one lungfish studied after 13 months of aestivation the plasma osmolarity was $650 \, \text{mOsm} \, l^{-1}$ (control value 234), 31.2% being due to urea but with $184 \, \text{mmol} \, l^{-1}$ sodium (control 101) and $45 \, \text{mmol} \, l^{-1}$ sulphate (control 2.6).

When lungfish cocoons are immersed in water the fish emerge and increase greatly in size as they swell osmotically. Urea diffuses out rapidly across the gills but it is several days before the animal reaches a steady state. Urine production begins at a very high rate to eliminate the excess water entering. This diuresis can be prevented by hypophysectomy after

Figure 6.2 Changes in urine flow rate, plasma and urine osmolarities and blood urea concentrations during artificially-induced aestivation in an African lungfish, *Protopterus aethiopicus* (redrawn and modified from Delaney *et al.*, 1977).

which the fish will die. Posterior pituitary hormones have an extremely potent diuretic action in *Protopterus*, no indication of the teleostean anti-diuretic effect being present.

Amphibian water balance

Lungfish cannot be said to be terrestrial animals, merely being able to tolerate limited periods without water, but they represent an interesting stage in the evolution of independence from the aquatic environment. Further steps towards a truly terrestrial existence are illustrated by various members of the Amphibia.

Although the adults of many Amphibia are terrestrial, very few species have become completely independent of the aquatic environment—all develop from eggs which must be laid and fertilised in water to hatch and undergo a larval stage there before metamorphosing into aerial forms. A few species have evolved ways of utilising very small amounts of water for the early stages in development or even rely entirely on fluid produced by the mother. Our interest in this chapter is, however, confined to the way Amphibia have solved the problems associated with emergence from water following metamorphosis.

Many studies have been carried out on osmoregulation in the Anura (frogs and toads). Although these have lungs and also breathe through the lining of the buccal cavity, most oxygen uptake is through the skin, which is kept moist by the watery secretions of the many cutaneous mucous glands. Water evaporates from the skin at about the same rate as from a free water surface, so in many species life is an alternating sequence of dehydration followed by return to water and rehydration. Tolerance of water loss is very well developed; many species can lose up to 50% of their body water before they die. This leads to an increased body fluid osmolality, but in the hydrated state amphibians differ from all other vertebrates in having a higher body water content (about 80% of body weight compared to 70%), with most of the extra water in the extracellular fluid, which is usually of lower osmolality (around $200\,\text{mOsm kg}^{-1}$) and sodium concentration (around $100\,\text{mmol l}^{-1}$). This means that water loss does not lead to excessively high solute concentrations. The bladder is very large and functions as a water store. In spite of all these adaptations water loss is so rapid in dry air that most frogs avoid exposure to it. This they do successfully even in arid conditions such as the Australian outback, where frogs survive by burrowing into the soil and aestivating, emerging only on the rare occasions when rain falls to breed in temporary pools. The eggs and larvae develop very rapidly to metamorphose before the pools dry up, and then burrow into the mud to await the next rains. The large volumes of water stored in the bladder during aestivation may play a role in the osmoregulatory adaptation of another species, *Homo sapiens*, to desert life. Aborigines short of water need only dig up a few frogs and squeeze the fluid from their bladders to obtain what is, in fact, very pure water.

The mechanisms of rehydration in Amphibia have been the subject of much research, as their skins and bladders make excellent experimental material for the study of water and salt movement across epithelia (see chapter 1). Water is absorbed passively through the skin, mainly that of the ventral surface, as the animals sit in pools of fresh water or even, in some species, by merely pressing the abdomen against moist soil. As the degree of dehydration and the body fluid osmolality increase, the hormone arginine vasotocin (AVT) is released from the neurohypophysis. This increases the osmotic permeability of the skin (see Table 4.3) and hence the rate of water uptake. It has the same effect on the bladder wall, causing reabsorption of the stored urine. At the same time AVT drastically reduces urine production by both decreasing GFR and increasing tubular water reabsorption.

Active uptake of salts accompanies water absorption by both skin and bladder, helping to prevent excessive dilution of the extracellular fluid. Bladder urine is very dilute in any case as a result of sodium and chloride reabsorption by the distal tubules, but further active uptake occurs across the bladder wall. Salt uptake is stimulated by AVT, but another factor is involved—reduced extracellular fluid sodium concentrations lead to increased release of the adrenal hormone aldosterone which stimulates sodium transport (chapter 1).

CHAPTER SEVEN

PROBLEMS OF WATER SHORTAGE

THE ANIMALS DISCUSSED IN THE PREVIOUS CHAPTER HAVE SUCCESSFULLY made the transition from an aquatic to a terrestrial existence, but are all dependent in some way on the presence of water. Some require access to liquid water, either at frequent intervals or just at one stage in their life cycle; others need to live in water-saturated air. In this chapter we will consider those animals which not only survive but thrive in dry air which would cause rapid desiccation in animals lacking their special adaptations. In the most extreme environments—the hot dry deserts—relatively few groups of animals are completely at home, mainly members of the Arachnida, Insecta, Reptilia and Mammalia. When we think of life in the desert it is animals like scorpions, locusts, rattlesnakes and camels which first come to mind, although representatives of many other groups can adapt to life in arid climates. Desert frogs were mentioned in the last chapter; African frogs of the genus *Chiromantis* have evolved relatively waterproof skins (a marked contrast to the great majority of the Amphibia) and phyllomedusine South American frogs possess glands which secrete waxy esters spread over their body surfaces by the wiping action of their limbs. Other rather surprising inhabitants of deserts include snails which can seal themselves inside their waterproof shells during the day.

Prevention of water loss

The first essential adaptation to dry conditions is limitation of water loss. The various sites of water loss will be considered separately.

The integument

Insects and arachnids, which may well have been the first, and which are certainly, in terms of number of species, the most successful terrestrial animals, have thin waterproof layers of waxes or greases on the outside of their cuticles. The molecules of grease or wax are arranged not randomly but in organised stacks which virtually prevent water loss. This organisation is disrupted if the animal is raised above a species-specific transition temperature. This disruption is irreversible in species with long-chain, hard-wax waterproofing but in species employing short-chain greases (e.g. the cockroach, *Periplaneta*) the impermeable layer re-forms on cooling. Reptile skin is dry and relatively impermeable but it is now clear that evaporative water loss through the skin may be of significance in some species, especially just after sloughing. Bird and mammalian skin is also substantially impermeable to water; although mammalian skin contains rather less keratin than is present in reptile skin there is little difference in the insensible water loss rate. (Loss of water in the form of sweat is a totally separate phenomenon and has nothing to do with skin permeability!)

Respiratory loss

Aerial respiration demands that a large surface area must be present at which oxygen can dissolve in the body fluids. The air in contact with this surface is inevitably saturated with water vapour, and mechanisms have evolved to limit loss of this vapour from the animal. This is difficult in those animals, such as amphibians, where the whole skin is a site of gaseous exchange, but terrestrial animals may have enclosed respiratory organs from which water loss can be limited. In all but the very smallest animals, where diffusion into and through the body is possible, oxygen is delivered to the tissues in one of two ways—directly, as by the insect tracheal system, or indirectly, as by the vertebrate system of lungs and a circulatory system.

How can water vapour loss be limited by enclosing the gas exchange surface within the body of the animal whilst allowing other gases—oxygen and carbon dioxide—to exchange with the atmosphere? Various strategies are employed. Some insects open the spiracles at the entrances to their tracheal system only intermittently. During the "open" periods a burst of CO_2 and water vapour is nevertheless lost (figure 7.1). A surprising finding was that oxygen uptake continued during the "closed" periods, showing that a slight degree of opening allowed air entry, but not carbon dioxide and water loss. This seems to be possible because, as the respiratory quotient is less than 1, more oxygen is consumed than carbon dioxide is produced and a slight negative pressure develops in the tracheae. Diffusion through the very small openings is restricted so the slight pressure gradient is maintained and the slow inflow of air prevents water vapour from diffusing out. Unfortunately, the tracheae must be opened at intervals to allow accumulated carbon dioxide to escape, and there is then some unavoidable loss of water vapour. Large amounts of carbon dioxide can be lost during the "open" periods however, because as soon as some diffuses out it is replaced from the tissues, where it has been stored as bicarbonate

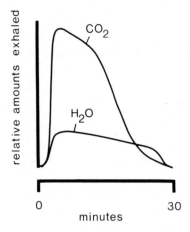

Figure 7.1 Relative amounts of CO_2 and H_2O exhaled during a "burst" of spiracle opening in a *Hylophora* pupa in diapause (redrawn and modified from Kanwisher, 1966).

ions during the inter-burst period, and this helps to maintain the concentration gradient for diffusion. During flight, when oxygen demand is greatly increased, the spiracles remain open and water loss is considerable, so it is advantageous for a desert insect to limit its flying activity during the heat of the day.

Limiting the time during which gas exchange between the atmosphere and the respiratory tract occurs is a strategy also employed by vertebrates. Rattlesnakes, for example, breathe only two or three times an hour. Oxygen uptake from the air in the lungs continues between breaths, but water vapour is lost only during the exhalations. Water loss amounts to as little as 0.4 % of the body weight per day in dry air at 23°C (more than 70 % of this is from the skin). Rattlesnakes can survive this loss for 2–3 months without any water, but would normally spend some time underground, where water loss is much less because of the higher humidities and lower temperatures of the burrows.

Endothermic animals, which use metabolically generated heat to regulate their body temperature, are at a disadvantage in hot arid conditions compared with ectothermic animals, which employ behavioural thermoregulation and have a much lower basal metabolic rate. Mammals and birds have high resting oxygen requirements and this is correlated with the possession of completely divided hearts, so that they always have to pump blood to the lungs at exactly the same rate as to the whole of the rest of the body. Reptiles can divert blood away from the lungs between breaths, and thus do not have to expose the blood to surfaces from which evaporation will occur if this is not necessary to meet respiratory requirements. Maintenance of a constant optimum body temperature by regulating the amount of solar radiation absorbed works very efficiently in hot dry surroundings. Whoever thinks that the thermoregulatory and circulatory systems of reptiles are more "primitive" than our own has obviously not tried to live without water in the desert! Animals which maintain their body temperature above that of their surroundings are at another disadvantage in dry climates. Air in the lungs is saturated with water vapour at body temperature and breathing out warm moist air results in large water losses. Some desert animals have special anatomical features to reduce the temperature of the exhaled air. The kangaroo rat, *Dipodomys*, has long narrow nasal passages with a large surface area. Inhaled air is gradually warmed as it passes through these, and is moistened by evaporation of water which helps to keep the nose cool. Exhaled air cools as it passes through the nasal passages, and water condenses. In fact a countercurrent system exists in which opposing flows are separated, not spatially but temporally. The air leaves the nose at close to ambient temperature and the further below body temperature this is the more water is conserved. To give an example, saturated air at 38°C contains 46 mg water per litre whereas at 25°C it contains only 23 mg per litre, so if the exhaled air is reduced to this temperature, respiratory water loss is halved. In other mammals, including man, air is exhaled at not much less than body temperature, and more water is lost. Some birds and reptiles have an additional source of fluid to evaporate into the inhaled air current—the nasal gland secretion (see chapter 8). The marine iguana, for example, has a depression just inside the external nares in which nasal gland outflow collects. The glands secrete a concentrated solution to conserve water, and it is further concentrated by evaporation with no expenditure of metabolic energy. Even in the absence of any cooling effect on the nose, the water which is added to the inhaled air replaces body water which would be lost in the saturated exhalant air.

Thermoregulatory loss

In hot climates, endotherms encounter further problems of water loss not associated with respiratory exchange. Air temperatures can easily exceed the body temperature of mammals (usually around 34–36°C), although birds, which have rather higher body core temperatures (40–42°C), are less affected. Mammals therefore risk sunstroke and, since an increase in body temperature of more than about 4°C is usually lethal, they must lose heat by evaporative cooling. Two avenues for this are open, sweating and panting. The relative importance of these two mechanisms varies greatly between mammalian species (man relies wholly on sweating, dogs on panting) but considerable quantities of water can be lost by either route. Sweating, because fluid containing sodium salts is secreted, does not cause the increase in blood concentration induced by panting, which liberates pure water. In this sense it is more efficient, but it can lead to salt depletion unless salts, as well as water, are replaced at intervals. Loss of water alone leads to increased extracellular fluid solute concentrations, causing water to move out of the cells. Loss of water plus sodium salts does not produce cellular dehydration but reduces the extracellular fluid volume. Panting is the only source of evaporative cooling in birds, though it is not quite the same mechanism as in mammals since evaporation takes place in the respiratory tract rather than from saliva in the mouth.

Neither panting nor sweating is a realistic mechanism for small endotherms, since their relatively high metabolic rates require high oxygen uptake which in turn demands high respiration rates. The margin left for panting is too small, and, in any case, any significant water loss will rapidly cause disastrous blood concentration. Consequently, most small mammals of all habitats do not have sweat glands, and desert species, such as the kangaroo rat, must avoid dangerous increases in core temperature by remaining in the sun for short periods only. The problems of being small as well as endothermic become most severe in small birds, especially humming birds. As well as having a higher core temperature, all birds have a rather higher metabolic rate than mammals and lose a minimum of about 5 % of the body weight by respiratory water loss each day (cf. 1 % per day in man). Smaller birds, unlike small mammals which can burrow or live in damp vegetation, may lose up to 40 % body weight daily by the same route. Humming birds have a further problem in that their mode of life and food capture demand an exceptionally high metabolic rate (up to 500 times that of a resting human) when flying. They are therefore limited to areas where the day temperature is not far from their core temperature so they do not need to lose much heat by evaporative cooling, nor is a great deal of energy required to maintain their core temperature when the environmental temperature drops somewhat. Problems arise for humming birds at night, when temperature decreases of 20°C or more may occur. To maintain a core temperature of more than 40°C against such a temperature gradient would require a tremendous expenditure of energy for a form with such a high surface area to volume ratio. This in turn would require a huge oxygen uptake with consequent increase in evaporative water loss. Such depletion of both metabolic and water stores is avoided because the birds go into a state of torpor, becoming ectothermic at night, which reduces their respiratory needs by up to 90% (Lasiewski, 1964).

Excretory loss

Faecal loss

Some water is always lost with the faeces, although water reabsorptive mechanisms exist in the alimentary canals of all animals. Perhaps the most efficient of these are found in some insects in which the rectum contains dry faecal pellets in air. Faecal water losses are very variable; carnivores or organisms subsisting largely upon dry plant materials such as seeds tend to produce small and/or dry faeces. On the other hand, species eating large quantities of fibrous vegetation can lose significant quantities of water in copious faeces. The donkey, *Equus asinus*, which is naturally a desert form, loses nearly twice as much water in the faeces as in the urine. Mechanisms of avoiding faecal water loss have attracted much less study than urinary mechanisms, except in the insects where the two processes are so closely related. Mention must be made of the faecal eating habits of the rabbit; eating moist pellets which pass through the water reabsorbing areas of the colon twice rather than the normal once is an interesting but apparently not widespread approach to balancing the water budget! Termites, however, are even more sophisticated in their use of faeces. Dry pellets are voided during the day, pick up moisture from dew at night, and are eaten to gain the excess water.

Urinary loss

Fluid loss from excretory organs must obviously be restricted as much as possible when water is in short supply. Although the way in which this has been achieved has been studied in many groups, convergent evolution has resulted in a few trends which allow convenient generalisations to be made. Many waste products have to be excreted, but only excess salts and nitrogenous wastes are likely to be present in such amounts as to affect water balance. Salt excretion will be dealt with in the next chapter, so here we will consider excretion of the end products of nitrogen metabolism only.

As we have seen in the last chapter, whereas aquatic animals can eliminate ammonia directly into the water, the transition to land has meant that relatively non-toxic nitrogenous compounds have to be synthesised and eliminated by the excretory organs. In the case of most amphibians, urea is the main nitrogenous waste product, except in aquatic forms which excrete ammonia. Many terrestrial animals, including amphibians and reptiles, are incapable of producing urine more concentrated than the body fluids. To eliminate 1 g of waste nitrogen in urine iso-osmotic with 300 mOsm kg^{-1} body fluids would require about 110 ml of urine if no other solutes were present. Many terrestrial animals, including insects, reptiles, birds and even a few amphibians, excrete most of their nitrogenous waste as urates (uric acid or its salts). The uric acid molecule contains twice as much nitrogen as the urea molecule, so in the example above only about 55 ml of urine would be required to eliminate 1 g of nitrogen in the form of dissolved uric acid. The main advantage of uric acid and its salts, however, is that they are relatively insoluble, and precipitation of urates in Malpighian or renal tubules prevents them from exerting osmotic effects. Thus a thick sludge can be excreted which contains large amounts of nitrogen, but is nevertheless not hypertonic to the body fluids. If uric acid is such an ideal waste product why is it not used by all animals? The answer seems to be that its synthesis requires a greater expenditure of energy, in the form of ATP, than that of urea, and where water is available urea or, if possible, ammonia is produced instead with no waste of metabolic energy. Table 7.1 shows how variable the nitrogenous excretory patterns may be within a single order of reptiles, the Chelonia (tortoises and turtles). The examples given are selected from a large number of analyses, the proportion of ammonia, urea and urate often changing dramatically within the same species under different conditions.

The mechanism of urinary excretion of urate (uricotely) has been greatly

Table 7.1 Nitrogenous excretion in Chelonia

Species	Habitat	% of urinary nitrogen		
		Ammonia	Urea	Urate
Chrysemys scripta	fresh water	79	17	4
Kinixys erosa	moist terrestrial	6	61	4
Testudo graeca	dry terrestrial	4	22	52
Gopherus berlandieri	desert	4	3	93

oversimplified in undergraduate texts in the past and the following points should be noted:

1. It is inaccurate to describe the excretory product as "uric acid". Urate salts make up much of the solid material in many cases, and uric acid, if present, exists as uric acid dihydrate.

2. Relatively recently it has been discovered that significantly large quantities of cations are present in the solid urinary pellets of some lizards (e.g. *Dipsosaurus*). All of the available evidence suggests that they are excreted as monobasic urate salts. The significance of this in terms of potassium excretion will be considered in the next chapter, but interestingly it has been found that ammonia can be excreted in solid form as ammonium urate without raising the osmotic concentration of the urine. Provided urinary water is reabsorbed in a post-renal structure (cloaca/colon/rectum/urinary bladder) large amounts of salts and nitrogen can be excreted in insoluble form without excessive water loss or the necessity for producing a hyperosmotic urine.

3. Kidneys are primarily designed for the transport of fluids, not solids. How then do they cope with the transport of urates which have low solubility and therefore precipitate easily? It is now apparent that true hard crystals of urate are normally only formed in cavities (cloaca/colon/bladder) distal to the ureters. It appears that urate in the kidneys is present either as a colloid or in the form of smooth-walled, non-crystalline, gel-like spheres ($2-10 \ \mu m$ in diameter) which slide easily along the ureters and past each other.

Insects secrete most of their nitrogenous waste as urate, or (in some species) as its breakdown products allantoin and allantoic acid, using a mechanism in which the digestive and excretory systems work in harness. The primary organs of excretion are the Malpighian tubules (which also occur in arachnids), thin-walled ectodermal structures with a blind (distal) end projecting into the haemolymph and opening into the junction between the midgut and the rectum (figure 7.2). The number of tubules varies between $2-150$ depending on species, but the total surface area is roughly constant in relation to the haemolymph volume. The epithelial cells secrete an almost protein-free fluid which is nearly iso-osmotic—in fact usually very slightly hyperosmotic—to the insect's haemolymph. Fluid secretion in distal parts of the tubules is dependent on active salt (usually potassium) transport. In some blood-sucking insects, which absorb large excesses of sodium ions, sodium replaces potassium as the main cation in the tubule fluid. Maddrell (1977) has suggested a model (figure 7.3) in which either sodium or potassium ions are pumped by an apical transport

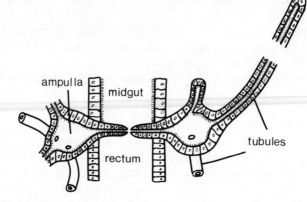

Figure 7.2 Semidiagrammatic representation of the form of Malpighian tubules in the cockroach, *Periplaneta*.

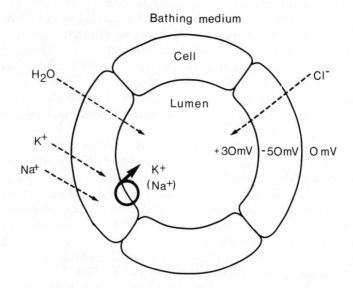

Figure 7.3 Ion movements and electrical potentials in Malpighian tubule cells. ↘ represents movement down electrochemical gradients, ⊘ ion pumps.

mechanism from the cytoplasm to the tubule lumen. Which ion will be pumped is determined by the permeability of the basal membranes. In most insects these are impermeable to sodium ions, but permeable to potassium ions, so that these diffuse passively into the cell. (The potassium concentration of insect haemolymph is usually higher than that of the extracellular fluids of other animals). They are then pumped actively into the lumen. Chloride ions follow passively down their electrochemical gradient and water follows down the slight osmotic gradient. The creation of local osmotic gradients between the apical microvilli or basal infoldings may assist this process.

The secreted fluid is then modified as it passes down the tubules. Solutes diffuse in from the haemolymph, and those of value to the animal are reabsorbed by active mechanisms, whilst unwanted substances, including urates, are actively secreted into the lumen. Final regulation is achieved in the rectum which modifies the composition of the fluid it receives in accordance with the requirements of the animal. The rectal glands, which vary enormously in their morphology between species, absorb water from the mixture of faeces and urine so that, in some instances, only solid matter containing very little water is excreted. The available data suggests that they do so as a result of the creation of local osmotic gradients (figure 7.4). A solute of some sort is pumped from the rectum lumen into the cells of the epithelium and thence into narrow intercellular spaces. This pumping creates local osmotic gradients directing water from the gut lumen into the luminal portion of the cell and from there to the intercellular spaces. Concentrated fluid from these spaces flows into larger intercellular spaces and thence via indentations of the basal membrane into the haemolymph. There is direct evidence for high osmolarities in the intercellular spaces, but there is considerable disagreement about the details of the mechanism. The system is particularly efficient in restricting water loss since the urine is concentrated and the faeces simultaneously dehydrated whilst waste in an insoluble nitrogenous form further conserves water.

Excretion of nitrogenous waste in reptiles and birds has many interesting parallels with the situation in insects. Although the fluid initially entering the renal tubules is formed by ultrafiltration at the glomeruli, urates are added by tubular secretion. Birds, unlike reptiles but in common with mammals, possess structures known as loops of Henle (see chapter 9) which enable them to produce hyperosmotic urine (but usually only up to about twice the plasma concentration, so this factor plays a minor role in water conservation compared with uricotely). Urine from the kidneys enters the cloaca and mixes with the faeces and may pass retrogradely into the rectum or colon. Here, as in insects, water is reabsorbed and the matter finally excreted is solid or semi-solid. Water uptake seems to be linked to active reabsorption of sodium. In many reptiles and birds the rectal-cloacal region plays an important role in salt

haemolymph

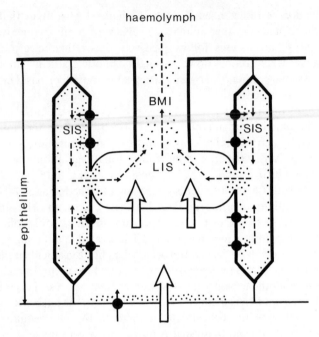

lumen of rectum

Figure 7.4 Model for rectal gland reabsorption of water in insects (simplified from Wall and Oschman, 1975). BMI = Basal membrane indentation, SIS = Small intercellular space, LIS = Large intercellular space, stippling (⠿) indicates high concentration of solute, arrow across solid circle (⊸) indicates active solute pump, open arrow (⇨) indicates osmotic water movement, dashed arrow (--→) represents bulk fluid movement, thin solid lines (—) indicate particularly water-permeable areas, thick solid lines (▬) indicate less water-permeable areas.

regulation and conservation. Those species faced with the problem of excreting excess salts have evolved alternative pathways, the salt glands (see chapter 8).

Mammals excrete nitrogenous waste as urea. The kidneys of desert mammals can produce strongly hyperosmotic urine containing very high urea concentrations (see chapter 9). Excretion of urea is wasteful not only from the point of view of water loss, but because the nitrogen it contains could be useful to the animal if resynthesised into protein. Unlike animals, bacteria can utilise urea or ammonia for this purpose, and ruminants take advantage of symbiotic bacteria in the rumen to perform this task. Urea diffuses into the rumen where it is metabolised and the bacterial protein produced is digested and absorbed as it passes down the alimentary canal.

Since food as well as water is in short supply in desert regions, this prevention of waste adds greatly to the survival capacity of ruminants such as the camel.

Some insects store uric acid in the fat bodies and may be able to reutilise it for protein synthesis. However, in animals with a very short life cycle, permanent storage of waste products in the body is a possible alternative to excretion. Adult Lepidoptera, for example, lay down uric acid crystals in the wings, and other insects do so in their cuticle, which is then shed at some time or times in the life cycle.

Replacement of lost water

Drinking
The evolution of drinking behaviour must have been of enormous importance in the colonisation of the land, although it is not the only way by which animals can obtain water. Amphibians for instance do not drink, relying instead on cutaneous water absorption (chapter 6), and uptake of water vapour from the atmosphere by some insects and arachnids will be discussed below. But most terrestrial animals rely on innate behaviour patterns, often showing circadian rhythmicity, to maintain a positive water balance. Regulation of body water content is then effected by controlling the amount of fluid eliminated by excretory organs.

If a water deficit is incurred, mechanisms exist to stimulate drinking behaviour. It is, of course, impossible to say whether other animals experience the same sensation of thirst as human beings do, but the term "thirst" is usually applied to the circumstances leading to the stimulation of drinking. Most studies have been carried out on mammals and it appears that the stimulus is loss of water from either the intracellular or extracellular fluid compartments, or both. Loss of water alone from the extracellular fluid leads to an increase in solute concentration which causes water to move out of the cells, producing intracellular dehydration. Loss of water plus sodium salts, e.g. in sweat, may however reduce the extracellular fluid volume without affecting its osmolarity, so no water will be lost from the cells.

The thirst centres in the brain are stimulated either by intracellular dehydration, detected when water is lost from osmoreceptor cells in the hypothalamus, or by a reduction in blood volume (hypovolemia). Blood volume changes are monitored by receptors in the thoracic veins, and possibly also in the brain. In addition, hypovolemia activates the renin-angiotensin system (chapter 9). Renin is released from the kidney and causes the production of the hormone angiotensin II in the blood. Angiotensin II, when injected into the ventricles of the brain, is the most potent dipsogen (agent which provokes drinking) known. It appears to act on two neurohaemal areas—the prefornical organ and the organ vasculosum of the lamina terminalis—possibly by constricting their blood supply and reinforcing their ability to respond to changes in blood volume.

A redundancy of mechanisms thus exists, any one of which will alone be sufficient to provoke drinking behaviour in the event of excessive water loss. Although these are vital to the survival of most terrestrial animals they are essentially emergency mechanisms. Many animals probably never experience thirst since they have evolved behaviour patterns which ensure that they always drink more than enough water.

The dry sensation in the throat which we associate with thirst is not itself the stimulus to drinking. It is merely a manifestation of a general dehydration which leads, amongst other things, to a reduction in saliva production. Mechanoreceptors in the alimentary canal monitor the amount of water drunk and ensure that drinking stops when sufficient water has been ingested, but because absorption from the alimentary canal takes time, any deficit will not be corrected until long after drinking stops. Many animals are most at risk from predators when visiting water sources, and any mechanism which limits the duration of drinking activity has obvious survival value.

Uptake of water vapour

Perhaps the most advanced way of living in the complete absence of liquid water is shown by some insects, mites and ticks which can absorb water vapour directly from the atmosphere. The most remarkable example so far discovered is the cigarette beetle, *Lasioderma serricorne*, which can reduce the relative humidity of the atmosphere around its body to 43%. A solution in equilibrium with such an atmosphere would have an osmotic pressure of almost 1200 atmospheres! The fact that an insect always reaches equilibrium with a fixed relative humidity and not a fixed water vapour pressure (temperature changes do not alter the figure) suggests that whatever mechanism is involved has similar properties to those of a hygroscopic substance. Saturated solutions of sodium chloride and potassium chloride reach equilibria with atmospheres of 75% and 83% relative humidity respectively, so recycling of these ions could not possibly create gradients of water activity sufficiently large to facilitate absorption. It is easy to envisage some hygroscopic substance which would have no difficulty in absorbing water to the required extent, but an active process would then be needed to release the water and make it available to the body. Metabolic energy is required for water uptake, which is reduced by metabolic inhibitors or during starvation, and stops on death of the animal. The process therefore appears (uniquely in the animal kingdom) to involve the active transport of water.

Elucidation of the mechanism of the "water pump", or even identification of its components, poses enormous practical problems and little progress has been made. The location of the sites of uptake have, however, been found in a few species. In several insects the rectum—or specialised structures associated with it, such as the anal sacs of the firebrat, *Thermobia domestica*—is involved. In some ticks, highly concentrated saliva, which equilibrates with air with a relative humidity of 80%, is secreted on to the mouthparts. After a time it is swallowed and presumably diluted and absorbed in the alimentary canal. The evolution of mechanisms of extracting water from relatively dry air has undoubtedly contributed to the ability of insects and arachnids to colonise arid environments and to their present position as by far the most successful terrestrial animals.

Water in food

In xeric areas, where water may be permanently scarce or only intermittently available, many animals rely on food as a water source; this is particularly true of insects feeding on plant juices, nectar or blood. Apart from this, some animals rely heavily upon metabolic water released by oxidative reactions such as

$$C_6H_{12}O_6 + 6O_2 \rightarrow 6CO_2 + 6H_2O.$$

Possibly the most famous example of such water economy is provided by the kangaroo rat studied by Schmidt-Nielsen (1964). Table 7.2 shows a

---

Table 7.2 Water economy of *Dipodomys* fed on dry barley

Credit	Debit
100 g barley gives 53.7 g oxidative water	1. 8.14 l of oxygen is needed to oxidise barley; *43.9 g* of water is lost by evaporation in breath.
	2. Metabolising the barley results in 3.17 g of urea which must be excreted; *13.5 g* of water is needed for this.
	3. Faecal loss corresponding to barley intake carries *2.5 g* water with it.
Total: $+53.7$ g	Total: -59.9 g

Deficit: 6.2 g

balance sheet for the metabolic water acquired when a desert rat eats 100 g of dry barley. Clearly, even the kangaroo rat cannot live by metabolic water alone, but the overall deficit is so small that dried barley equilibrated with air of only 20% relative humidity contains sufficient free water to bridge the gap.

Toleration of dehydration

Most animals, whether invertebrates or vertebrates, can tolerate only some 10% to 20% water loss, although there are a number of exceptions from a variety of groups. Amphibians which can survive the loss of half their body water were mentioned in the last chapter. One insect, the larva of the Nigerian chironomid midge *Polypedilum vanderplanki*, can survive the almost complete dehydration which results when the holes in rocks in which the larva lives dry out completely during the dry season. Insects have some advantage in withstanding dehydration in that haemolymph volume can be greatly reduced with no ill effects—it is not part of a closed circulatory system and is not involved in respiratory gas transport. For example, in cockroaches kept for 9 days without water, haemolymph volume declined from 134 to 53 μl with little change in osmolality. The haemolymph can thus function as a reserve of available water.

Evaporation of water is the only way the body temperature of an exposed animal can be kept below that of its surroundings. The human body is well-adapted to life in hot climates, as copious perspiration is possible (up to 15 l of sweat per day) so long as water is readily available. But in the absence of water, dehydration is rapid and there is a disproportionately large loss from the plasma. Eventually, sometimes after less than a day in very hot, dry conditions, the efficiency of the circulation decreases and body heat cannot be transferred to the skin quickly enough. An explosive rise in body temperature then occurs and the lethal limit is rapidly exceeded.

Integration of water balance

Survival under arid conditions depends on a combination of all the factors mentioned above. What better example of a well-integrated animal than the camel, whose ability to travel long distances in hot dry conditions without access to water has provoked speculation in the literature from the time of Aristotle? The camel may look as if it was designed by a committee but it is remarkably well-adapted to its desert habitat. Many aspects of its physiology were worked out about 25 years ago by Schmidt-Nielsen, but there is still plenty of scope for further work and speculation. The few studies on camel water balance have all been based on very few individual animals—quite understandably in view of the inconvenience and expense of working with such a large experimental subject.

The first question is, does the camel store water? After periods of dehydration, camels drink very rapidly exactly the right volume of water to restore their body fluid compartments to normal conditions. Since they never drink excess water in anticipation of future shortages, are there any water stores within the body which can be called upon in an emergency? A theory widespread at one time was that the fatty humps acted as metabolic water stores. This is a seductive hypothesis since 1 g of fat may be oxidised to yield 1.07 g of water (see above) but, as seen in Table 7.2, the fallacy in the argument is that the amount of water lost from the respiratory surfaces in taking up the oxygen required for fat oxidation at least matches, and probably exceeds, the metabolic water gain. However the hump(s) are useful energy stores since camels, like most other animals, restrict feeding when deprived of water. Like all ruminants the camel carries a large volume of water in the alimentary canal. Two objections have been raised to the idea that this might serve as an emergency water store. The first is that this volume is required for the digestion of cellulose by bacterial fermentation, but of course this does not apply if the animal is not feeding. The second objection is that this fluid is not water but an isotonic salt solution, so would be difficult to absorb. The latter objection is not valid since camels can survive and maintain their body weight when given nothing to drink but 4% sodium chloride solution. Such a solution is more concentrated than sea water, but it can be absorbed and the salts eliminated without difficulty by the kidney, which can produce urine with a chloride concentration of up to 1000 mmol l^{-1}. In fact three camels kept without water in the sun at Alice Springs in Australia for 9 days, with a maximum daily temperature of 41°C, were estimated to have lost 90% of their gut water and this would have accounted for about half the total water loss (MacFarlane, Morris and Howard, 1963). Whilst this water may be drawn on under extreme conditions for relatively short periods it is obviously a relatively short-term expedient. Camels can survive for only 15 days under the conditions described above. On the other hand, one camel was kept without water at a constant 22°C for 45 days, and survived quite happily with nothing to eat but dry hay which contained 12% water (Maloiy, 1972). The gut water cannot be depleted if digestion is to continue so other explanations must be sought for the camel's remarkable capabilities.

The key to the camel's success lies in its ability to limit water losses. Firstly, it produces very dry faeces and has very efficient kidneys capable of producing urine of over $3000\,\mathrm{mOsm\,l^{-1}}$—ten times the blood concentration. During dehydration, GFR decreases and tubular water reabsorption increases so that urine flow is greatly decreased. Circulating ADH levels increase as might be expected; so too do those of aldosterone, which presumably acts to increase salt, and consequently water, uptake from the gut. Low rates of urine flow are possible because as urea is recycled through the rumen bacteria, little need be excreted, and the same applies to salts since these are lost in the sweat. Urinary loss however is low in proportion to the total water loss from the animal in hot dry air, so there is limited scope for effecting large water savings by reducing urine flow. This is in contrast to the situation in small desert mammals, such as the kangaroo rat, where renal conservation of water makes a large contribution to the overall water economy of the animal since extrarenal losses are relatively low, thanks to their nocturnal and fossorial habits.

Secondly, camels use much less water in keeping cool under hot conditions than do other mammals. The first thing in the camel's favour is that it is large (about 500 kg) and hairy. Size and insulation ensure that the camel warms up relatively slowly during the day, thus staving off evaporative water loss processes, while at night (which may be cold in the desert) heat loss is slow and therefore energy reserves, oxygen and respiratory water are not wasted in maintaining body temperature. Also, sweating is delayed on hot days because dehydrated camels allow their core body temperatures to fluctuate (see figure 7.5). This unusual adaptation allows the camel to postpone sweating until the core body temperature has risen by about 6°C. Such a temperature rise in a 500 kg camel results in a storage of about 2500 kcal. To dissipate this heat load would require 4 l of water to be evaporated either in sweat or in the respiratory tract. Instead, the camel loses this heat by conduction and convection at night. It has also recently been demonstrated that the nasal passages of camels function as heat exchangers to limit respiratory water loss, like those of kangaroo rats. As well as reducing the temperature of the exhaled air, the nasal passages reduce its relative humidity to 75%. During the day, evaporation helps to cool blood flowing to the brain (Schmidt-Nielsen *et al.*, 1980).

Camels also tolerate dehydration much better than other mammals; they are still healthy after losing 25–30% of their body water (an adult human would be *in extremis* after losing 10–15%). Apart from the possibility of using gut water in an emergency, the key to ability to

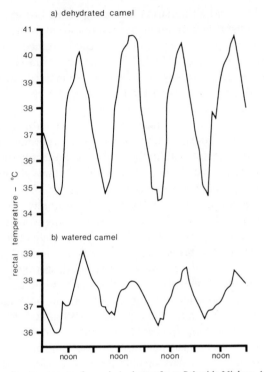

Figure 7.5 Rectal temperatures of camels (redrawn from Schmidt-Nielsen, 1964).

withstand such losses seems to be that the decrease in plasma volume is proportionately less than the total decrease in body water, and circulatory efficiency is not impaired. The accuracy and speed with which water deficits can be replaced is remarkable. As much as 100 l can be drunk in 10 minutes and the intake matches the deficit to better than 1 %! In contrast, a dehydrated human takes many hours to regain his original water balance.

Reproduction in the absence of water

The early developmental stages of animals can exist only in an aqueous environment, until the various organs required for terrestrial life have begun to function. In many amphibious animals, such as the desert frogs mentioned in the last chapter, the adults are well-adapted to life in arid environments but have to return to water to breed, the eggs developing into aquatic larvae which undergo metamorphosis before they can emerge on to land. To avoid an aquatic phase in their life history, animals must provide their offspring with an artificial aquatic environment in which to start life. Two ways of doing this have each evolved several times in different invertebrate and vertebrate groups; these are viviparity and the laying of cleidoic (closed) eggs. In viviparous animals the embryo develops in a fluid-filled compartment within the maternal body, which handles all exchange of materials with the environment. It has often been assumed that the embryo therefore faces no osmoregulatory problems; the mammalian foetus however exists in a slightly hypotonic amniotic fluid, and it has been suggested that the hormone prolactin (the "freshwater hormone" of fish), which is present in large amounts during pregnancy, might be involved in foetal osmoregulation.

Cleidoic eggs are laid by reptiles, birds, monotreme mammals, insects and gastropod molluscs. All are essentially water-impermeable boxes containing an aquatic environment for the developing embryo. Eggs may contain water bound to proteins, such as egg albumin, or lipids which can act as a source of metabolic water. Birds and insect egg shells are very impermeable to water (nevertheless almost half the water content of a bird egg is lost by evaporation during the course of incubation); those of reptiles (especially chelonians) and molluscs less so and they usually have to lay their eggs in damp places. Eggs contain stored protein which is broken down into its constituent amino acids and resynthesised by the embryo for use in growth. Any surplus amino acids can be broken down to provide energy, and it would appear to be advantageous to provide an excess of these essential building blocks for embryonic growth since in fish and frog eggs 70% of the embryo's energy requirements are met from protein catabolism and only 30% from fat. The ammonia produced from unwanted protein nitrogen can easily diffuse out of aquatic eggs, but in a cleidoic egg would rapidly reach toxic levels. Protein catabolism is therefore greatly reduced—in a chicken egg only 6% of the energy used by the embryo is derived from protein and 80% from fat—but some protein breakdown is inevitable. In reptiles which lay eggs in damp substrates it is possible for the embryo to excrete unwanted nitrogen as urea, which passes through the somewhat permeable egg shell. In most eggs, however, the excretory product is urate which precipitates and is stored until hatching. The insoluble crystals produced by purine synthesis are harmless to the embryo and take up little space. Incidentally, similar storage of water occurs in lepidopteran pupae (a sort of cleidoic egg!); the accumulated nitrogenous waste is voided shortly after the imago emerges. Urea, on the other hand, which would remain in solution and eventually reach toxic levels, is a more convenient waste product in viviparous animals because of its solubility—it can diffuse across the placenta and be eliminated by the mother. The evolution of uricotely in response to problems of water shortage in adults probably made the evolution of the cleidoic egg possible, thus marking the final step in the achievement of complete independence from the aquatic environment for many animals.

CHAPTER EIGHT

SALT REGULATION

Shortage of salts

Aquatic animals absorb salts from water, and amphibians continue to do this when they return to water, but animals which are wholly terrestrial have to rely on dietary salt intake, coupled with mechanisms for preventing loss of salts which are required and for excretion of excesses. Carnivorous animals can easily acquire the salts they need from their food since it is likely to have a similar composition to their own bodies, and their only problems are likely to be excessive intake of salts, for example in mammals which feed on marine invertebrates. Herbivorous animals may however experience difficulties because plants, whilst always rich in potassium, usually contain very little sodium. As sodium is the main cation in the extracellular fluid of most animals, ensuring an adequate uptake is important. Some groups of animals have evolved extracellular fluids with reduced sodium concentrations—amongst the insects, primitive orders have high sodium haemolymph, whilst in the more advanced orders sodium may be replaced to a greater or lesser extent by magnesium or potassium ions or by organic solutes (mainly amino acids). Herbivorous insects may have either a high or low ratio of potassium to sodium ions in the haemolymph, but in carnivorous insects the ratio is always low.

Soils, and therefore plants, deficient in sodium tend to be found in continental areas far removed from oceanic influences, particularly where high rainfall leaches salts out of the soil. Many animals show a specific appetite for sodium salts. Experiments were carried out in Australia on rabbits feeding on sodium deficient pasture. Wooden pegs soaked in various salt solutions were placed in the ground and it was found that those soaked in sodium chloride and, to a lesser extent, sodium bicarbonate were rapidly chewed away by the rabbits whilst those soaked in other salts were ignored (Blair-West *et al.*, 1968). Mammalian tongues have specific taste receptors for salt, illustrating that it is important to terrestrial animals to recognise sodium salts. Tropical herbivores congregate in large numbers at salt licks and access to salt must have some survival value to make it worth the consequent risks of vulnerability to predators. Ruminants have the most severe problems of sodium availability since they need to secrete large amounts in the saliva to produce the necessary volumes of rumen fluid in which bacterial fermentation of cellulose occurs. Animals which cool their bodies by sweating also lose large amounts of sodium in the sweat, in contrast to those which employ the alternative strategy of panting, in which water only is lost.

Sodium ions in both saliva and sweat can be partly replaced by potassium ions under the influence of the hormone aldosterone produced during sodium depletion (see chapter 9). There is some evidence that angiotensin stimulates sodium appetite, although this is still controversial. The renin–angiotensin–aldosterone system appears to have evolved to regulate sodium balance in terrestrial vertebrates, and mineralocorticoids can also stimulate salt appetite. However, the main stimulus for sodium appetite is the reduction in blood volume which results from sodium depletion (chapter 7).

The question of whether humans have a physiological need for sodium chloride which causes them to add salt to their food has been much discussed (Hollenberg, 1980). Literature on the importance of salt extends from Imperial Chinese physicians writing 4500 years ago, to the writings of Mahatma Gandhi. The importance the Romans attached to salt is illustrated by the derivation of the word "salary" from the Latin word for salt. There is trade in salt between tribes in the interior of New Guinea (who otherwise have no contact with the outside world) and tribes living on the coast. However salt is not added to food in all cultures, and most people eat far more sodium chloride than they need as it seems we rapidly become habituated to a high-salt diet, which may begin with salty baby foods. Reducing sodium intake may be beneficial and can be effective in the treatment of high blood pressure. The mechanism behind this effect is obscure; one hypothesis (Blaustein, 1978; de Wardener and MacGregor, 1980) is that a so-far undiscovered "sodium excreting hormone" (see chapter 9) is released in response to excess sodium intake and that this affects sodium–calcium exchange mechanisms in the cell membranes of vascular smooth muscle cells. This increases the intracellular ionic calcium concentrations which increases the resting tone of the muscles causing vasoconstriction and hypertension. This is all very speculative and illustrates how little we know about some physiological mechanisms. A more obvious explanation is that a low-sodium diet causes sodium depletion and a reduction in extracellular fluid volume including the blood volume, therefore reducing the blood pressure. However, normal salt balance can be maintained with an extremely low intake. It seems that the human body normally has no problems in maintaining body sodium levels on a normal dietary intake by regulating the amounts excreted, and this probably applies to most terrestrial animals.

Excess salts

Problems of excess absorption of salts in the diet and of water shortage are closely related since water is required for salt excretion. The ability to concentrate the urine reduces water loss and makes a high salt and low water intake possible. The mammalian kidney (chapter 9) and the insect rectum (chapter 7) are two organs which confer this capability on their owners. Uricotelic animals are able to excrete cations as insoluble urates (and significant quantities of soluble salts can be trapped in urate crystals) increasing salt excretion without necessitating the production of hypertonic urine. Birds can produce a hypertonic urine but the existence of solute-linked cloacal water reabsorption shows that, in some species at least, water retention has a higher priority than urinary salt excretion. The discovery by Schmidt-Nielsen and co-workers that the avian nasal gland is capable of secreting a concentrated salt solution has stimulated much research into extrarenal salt excretory organs in reptiles and birds.

Particular problems of salt loading are encountered by secondarily marine organisms and by herbivores living in arid regions. In the former, the excess consists largely of sodium and in the latter, of potassium salts; for convenience, regulation of these two ions will be considered separately.

A. Excess potassium

Desert animals, particularly reptiles, have been much studied. One non-desert species known to be particularly susceptible to accumulation of excess potassium is the Galapagos marine iguana, *Amblyrhynchus cristatus*. This large lizard spends much of its time in or near the sea and feeds largely upon potassium-rich seaweed, so is susceptible to entry of excess potassium and sodium ions. Tolerance of high blood potassium concentrations (hyperkalemia) is not possible for any of these potassium-stressed animals. Heightened extracellular potassium initially makes cells more excitable, but at very high levels all excitability is abolished. Reptiles such as the desert tortoise, *Gopherus agassizi*, rely on urinary excretion of potassium as precipitated urate salts. Others extrude potassium from salt glands which, unlike those of marine birds and reptiles which will be discussed in the next section, secrete a fluid which is rich in potassium rather than sodium ions. (There is also a slight tendency for bicarbonate to replace chloride as the dominant anion.) These salt glands are extremely important to herbivorous reptiles; *Dipsosaurus dorsalis* excretes 43 % of ingested potassium, 49 % of ingested sodium and 93 % of ingested chloride via its nasal glands, and *Amblyrhynchus cristatus* secretes a fluid in which both sodium and potassium concentrations are high ($1434 \, \text{mmol} \, \text{l}^{-1} \, \text{Na}^+$ and $235 \, \text{mmol} \, \text{l}^{-1} \, \text{K}^+$ has been recorded). The ostrich also secretes a potassium-rich solution from its nasal glands.

B. *Excess sodium*

Vertebrates living in or near the sea tend to gain salts from their environment because their body fluids are hypo-ionic to sea water. The resulting problems of teleosts and elasmobranchs have been dealt with elsewhere (chapters 3 and 5); here we are concerned with animals of terrestrial ancestry which have reinvaded marine and estuarine habitats—marine mammals (e.g. whales, seals), birds and reptiles (marine turtles, sea snakes, estuarine crocodiles, brackish-water lizards and chelonians).

Marine mammals maintain blood concentrations slightly higher than their terrestrial relatives; dolphin plasma for example is about $420 \, \text{mOsm} \, \text{l}^{-1}$ whereas that of terrestrial mammals is around $300 \, \text{mOsm} \, \text{l}^{-1}$. The animal is still markedly hypo-osmotic to sea water. However, marine mammals have salt- and water-impermeable skins and do not have sweat glands, which would be useless in water anyway, and they always breathe air of high humidity. Consequently urinary and faecal losses make up about 90 % of the total, relatively small, water loss and they can balance their water budgets with the free and metabolic water of their food; no marine mammal is known to drink sea water. This still leaves the problem of the salt (mainly sodium chloride) intake with food. Fish-eating seals and toothed whales have less of a problem in dietary salt intake since their prey has a body fluid/tissue osmotic and salt concentration well below that of sea water. On the other hand, walrus and krill-eating baleen whales consume invertebrate prey which imposes a greater loading, especially of chloride ions. However, the *mean* body fluid chloride concentration of a marine osmoconformer is still much less than that of sea water and baleen whales are able to produce urine with a high enough chloride concentration to be able, in theory at least, to cope with the problems produced by drinking sea water.

Marine birds and reptiles have problems similar to marine mammals but lack the ability to produce urine sufficiently concentrated to eliminate the excess load of ingested sodium chloride. A number of reptiles, such as the estuarine crocodile, *Crocodylus porosus*, and the soft-bellied turtle, *Trionyx triunguis*, simply tolerate a gradual increase in body sodium content. In crocodiles it is not entirely clear where the sodium goes to, since plasma levels are not elevated. In *T. triunguis* however, as in several desert lizards, the blood sodium rises (hypernatremia) and the animal must find a source of fresh water before the level becomes critical. The lizard *Amphibolurus ornatus* has a normal plasma sodium concentration of $150 \, \text{mmol} \, l^{-1}$, but during summer droughts this may rise above $240 \, \text{mmol} \, l^{-1}$. Not surprisingly, this Australian agamid drinks heartily when the rainy season returns.

Salt glands in reptiles and birds

Marine birds do not tolerate hypernatremia and although they may have an advantage over marine mammals in that they can sometimes obtain fresh water by flying inland, they also lose considerable amounts of water by evaporation from the respiratory tract. An extrarenal route for the removal of salts without excessive water loss is therefore imperative and, as in marine and desert reptiles, it is provided by the salt glands.

Salt glands may be present in birds, lizards, sea snakes, chelonians and possibly crocodiles. Although the glands of all species are similar in structure and function they are not homologous, having evolved from different glands in the head. In birds and lizards the salt glands are modified nasal glands, in turtles and terrapins they are modified lachrymal glands and in sea snakes sublingual salivary glands are employed (figure 8.1). It is possible that the Harderian glands in the orbits of estuarine crocodiles may function as salt glands but so far only very young specimens have been studied—no-one has tried to collect "crocodile tears" from an adult.

All salt glands consist of lobules containing thick-walled tubules emptying into central canals which eventually join ducts draining into the nasal passages, the orbit or the mouth. In each lobule (figure 8.2) the blood capillaries are arranged to form a countercurrent system with the secretory tubules. This would appear to be necessary because although salt gland blood flow increases dramatically during secretion (in one goose up to $26.9 \, \text{ml} \, g^{-1} \, \text{min}^{-1}$) the rate of excretion of ions is so high that a large proportion of the sodium and chloride passing through the blood system

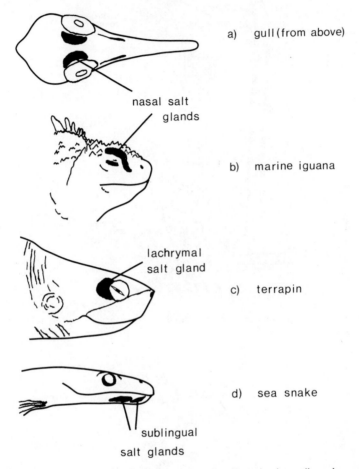

a)   gull (from above)

nasal salt
      glands

b)   marine iguana

lachrymal
salt gland

c)   terrapin

d)   sea snake

sublingual
salt glands

Figure 8.1 Location of salt glands in birds and reptiles. (Location in reptiles redrawn from Dunson, 1976.)

is extracted (up to 80% of the blood chloride in another goose). Because plasma potassium concentrations are much lower than those of sodium and chloride, enormous blood flows (up to $60\,ml\,g^{-1}\,min^{-1}$) would be required to deliver enough potassium to keep up with the known secretory rates of desert reptiles or ostriches.

Salt glands in marine birds and reptiles usually secrete a solution which is almost entirely sodium chloride of concentration greater than sea water, sometimes over $1000\,mmol\,l^{-1}$. Most studies have been carried out on

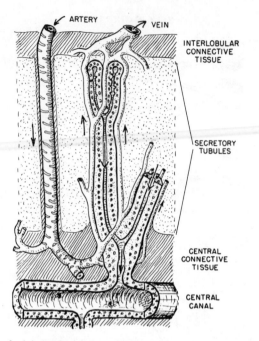

Figure 8.2 Part of a lobule of a herring gull salt gland to show a secretory tubule and its blood supply (from Fänge *et al.*, 1958).

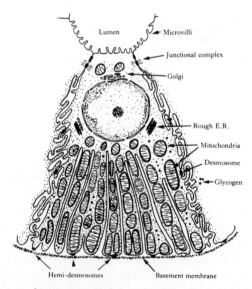

Figure 8.3 Structure of a secretory cell of a salt gland (from Peaker and Linsell, 1975).

birds, usually for convenience on domestic ducks and geese and not necessarily on those species with the most efficient salt glands. Secretion is not continuous but can be initiated by sodium loading, which is thought to stimulate osmoreceptors in or near the heart. These send impulses to the brain along the vagus nerves causing signals to be sent along the secretory nerves to the nasal glands, which secrete at a high rate until the salt load has been eliminated. Decreased extracellular fluid volume may produce elevation of the osmotic threshold (Kaul and Hammel, 1979).

The secretory cells are structurally remarkably similar to other salt-excreting cells—the chloride cells of marine teleosts and the cells of the elasmobranch rectal gland (figure 8.3). All are characterised by consider-able infolding of the baso-lateral membranes (in chloride cells and bird nasal gland cells it is the basal membranes, and in reptile salt gland and elasmobranch rectal gland cells the lateral membranes which are infolded). The membrane system which fills the basal part of the cytoplasm is very rich in $Na^+ + K^+$-activated ATPase, and is always interspersed

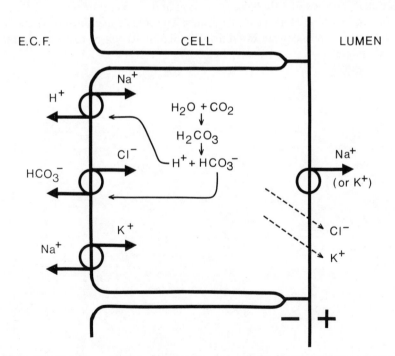

Figure 8.4 Possible mechanism of ion secretion by a salt gland cell. Dotted lines represent movement down electrochemical gradients.

with very large numbers of mitochondria. The student of comparative physiology might immediately suspect that some common salt-secretory mechanism must be associated with these remarkable structural similarities, but on present evidence this appears not to be the case. Most workers have concluded that active transport of either sodium or potassium ions across the apical membrane of the salt gland cells is the driving force for fluid secretion. Entry of sodium or potassium across the basal membrane may either be by passive diffusion or by chloride/bicarbonate and sodium/proton exchange mechanisms (figure 8.4).

It must be emphasised that most terrestrial birds, including many living in arid regions without regular access to free water, do not possess functional salt glands. Avian kidneys contain a mixture of "mammalian-type" nephrons, with loops of Henle, and "reptilian-type" nephrons without loops. The higher the proportion of the former, the greater the maximum urine osmolality, particularly as during conditions of water shortage, reptilian-type nephrons can cease filtering. Urine to plasma osmolality ratios of 2 to 2.5 are common and, as mentioned above, much salt can be excreted as insoluble urates. The role of the loops of Henle in the production of concentrated urine in mammals will be discussed in the next chapter.

CHAPTER NINE

THE MAMMALIAN KIDNEY

HOMER SMITH, IN HIS CLASSIC BOOK *FROM FISH TO PHILOSOPHER* TRACED THE whole evolution of the vertebrates in terms of their kidneys, but one need not take this extreme position to recognise the unique importance of kidneys in controlling with great accuracy the composition of the "internal environment" of mammals and thus making all the processes of life possible. The kidneys are not the only organs closely involved in mammalian osmoregulation—they would have nothing to regulate if solutes and water did not enter the body through the alimentary canal, and other organs play an essential, though negative, role in limiting losses of salts and water. But in contrast to other vertebrates in which a variety of organs (such as avian and reptilian salt glands, amphibian skin, fish gills) are active in the regulation of body fluid composition, the mammalian kidney combines the functions of all these structures. Mammalian species have colonised a wide range of extreme environments, from the seas to arid deserts, but can survive in a particular habitat only if they have a kidney functioning to maintain the correct internal environment.

A second reason for considering the mammalian kidney in some detail is the large amount of information available. Several thousand research papers are published each year on the subject; far more than on all other aspects of osmoregulation put together. The impetus for this great research effort is obviously the often tragic consequences of kidney disease, and although spectacular advances in medical technology have been made (for instance the introduction of "artificial kidneys" and renal transplantation), there are still many important areas of renal physiology which we do not understand. The vast amount of data which is accumulating, far from making it simpler to understand how the kidney works, raises so many new problems that sometimes established ideas are completely overthrown and it is impossible for textbooks to keep up to date with current thinking. In any case the functioning of the kidney is so complex that simple models may be inadequate to explain it. No chapter on the subject can therefore be a definitive account of "how the mammalian kidney works" but here we will attempt to provide some insight into current thinking and to illustrate principles which are of great importance in the study of osmoregulation generally.

Structure and function

It is important to remember that when medical physiologists refer to "the mammalian kidney" they are interested mainly in interpreting human kidney function in terms of observations on the laboratory rat. A few other species have been studied in detail, including the dog and the rabbit, but there is a tendency to discount work on species showing interesting differences from "typical" (i.e. human-like) renal structure and function, e.g. the North African sand rat, *Psammomys obesus*.

As Homer Smith was fond of pointing out, we have the type of kidneys we do, because our freshwater ancestors hundreds of millions of years ago needed a filtering apparatus to eliminate excess water. They embarked on an evolutionary trend which probably took as its starting point something like the primitive kidney of the present day *Myxine*, with its few segmentally-arranged glomeruli, and culminated in organs like the human kidney with its one million glomeruli. A kidney in which all extracellular fluid solutes smaller than proteins are removed from the circulation, necessitating the active reabsorption of all substances needed by the body, might seem inefficient, but without such a system we would have to be very careful what chemicals we allowed into our bodies. Specific secretory mechanisms would be required to eliminate each of them. Such mechanisms would probably have evolved to remove unwanted substances in our natural diet, but on the whole modern man is probably very fortunate to have evolved from freshwater animals!

The principles of renal physiology have been discussed in chapter 4 in relation to the lamprey kidney. The structure and function of the glomerulus has changed remarkably little from fish to man; in fact clinicians have been said to comment (on seeing an electron micrograph of a salmon kidney) that the patient has a perfectly normal healthy glomerulus! What *has* changed is the structure and function of the renal tubules. Birds and mammals differ from lower vertebrates in the possession of an additional segment between the proximal and distal tubule—the loop of Henle. Possession of such a loop is correlated with the ability to produce urine which is hyper-osmotic to the body fluids. Comparative studies on the kidneys of mammals from a wide range of habitats show that the maximum concentrating ability of the kidney is correlated with the length of the loop of Henle. For example the freshwater beaver, which never experiences water shortage, has very short loops and can produce urine of no more than $520\,\mathrm{mOsm\,l^{-1}}$. The human kidney contains a mixture of long and short loops and the maximum urinary osmotic concentration is $1400\,\mathrm{mOsm\,l^{-1}}$, whereas in the Australian desert hopping mouse all the loops are very long and the urine of up to $9400\,\mathrm{mOsm\,l^{-1}}$ can be produced. Since anatomical details are crucial to the functioning of the kidney the structure of the tubules and blood vessels must be described before the functions of the different parts of the nephron are considered.

Figure 9.1 is a diagrammatic representation of a segment of the kidney of a mammal, such as rat or man, in which there is a mixture of nephrons having long and short loops of Henle. The glomeruli and the convoluted proximal and distal tubules with their blood supply make up the cortex of the kidney. The medulla consists of the loops of Henle and collecting ducts and the vasa recta blood vessels. The long loops of Henle all come from juxta-medullary glomeruli in the inner cortex. The renal artery sends branches to the cortex only and each branch finally terminates in an afferent arteriole to a glomerulus. The peritubular capillaries contain only post-glomerular blood from efferent arterioles, the vasa recta blood vessels originating from the efferent arterioles of juxta-medullary glomeruli. The glomerulus, strictly speaking, consists of the bundle of capillary loops between the afferent and efferent arterioles which are surrounded by the Bowman's capsule. The Bowman's space, into which fluid passes by ultra-filtration (see chapter 4) is continuous with the lumen of the proximal convoluted tubule. The final straight part of the proximal tubule passes down to the inner cortex and leads into the loop of Henle. The final part of the ascending loop of Henle, which has a much thicker epithelial cell layer, connects with the distal tubule, which always returns to its own glomerulus before passing down to the medulla and joining the collecting duct. Collecting ducts from different nephrons join as they pass down to the papilla to deliver urine to the pelvis of the kidney which is drained by the ureter to the bladder where the urine is stored.

Some idea of the importance of the kidneys in the constant regulation of the extracellular fluid can be gauged from the fact that although the human kidneys only constitute about 0.5% of the body weight they receive 25% of the blood flow of the body at rest. This means that all the blood in the body passes through the kidneys every 20 to 25 minutes and since the blood plasma is constantly exchanging with the interstitial fluid across the

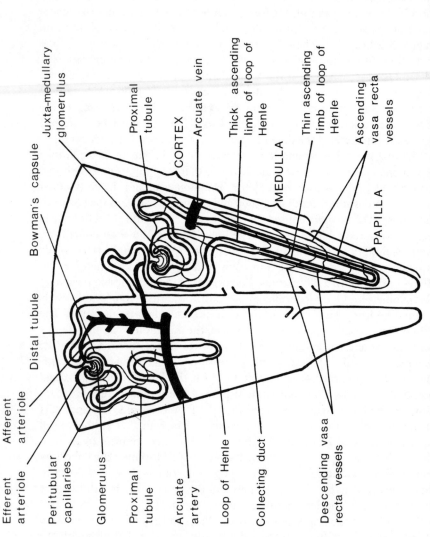

Efferent arteriole

Afferent arteriole

Peritubular capillaries

Distal tubule

Bowman's capsule

Juxta-medullary glomerulus

Proximal tubule

CORTEX

Arcuate vein

Thick ascending limb of loop of Henle

MEDULLA

Thin ascending limb of loop of Henle

Glomerulus

Proximal tubule

Arcuate artery

Loop of Henle

Collecting duct

PAPILLA

Ascending vasa recta vessels

Descending vasa recta vessels

Figure 9.1 Diagram to illustrate the structure of a segment of a mammalian kidney. Two types of nephrons are shown—a juxtamedullary nephron with a long loop of Henle and a cortical nephron with a short loop (based on diagrams in Moffat, 1975).

walls of the capillary beds and via the lymphatic system it means that all the extracellular fluid passes through the kidneys every few hours. In a 70 kg person, renal plasma flow (measured as clearance of p-amino hippuric acid, C_{PAH}, see chapter 4) is always about $700 \, \mathrm{ml \, min^{-1}}$ and glomerular filtration rate or GFR (measured as clearance of inulin, C_{inulin}) about $125 \, \mathrm{ml \, min^{-1}}$. Filtration fraction ($C_{PAH}/C_{inulin}$) is therefore 0.18. In other words about 1/5th of the plasma passing through the kidney, or about 180 l per day, is actually filtered—a volume equal to 4 times the total body water. Under normal circumstances more than 99 % of the fluid filtered is reabsorbed, so it will be appreciated that the kidneys are able constantly and precisely to control the composition of the body fluids.

Blood flow to the human or rat kidney does not increase if blood pressure is raised and cardiac output is increased. This phenomenon is known as autoregulation of the renal blood flow, and it is an active process involving changes in the tone of vascular smooth muscle, although its details are poorly understood. In contrast to the situation in the lamprey kidney, GFR remains remarkably constant at all times, irrespective of the state of water balance of the animal, and changes in urine flow are a result of changes in the amount of fluid reabsorbed by the tubules. The filtration rate of individual glomeruli is also precisely regulated. (In some mammals, such as the dog, GFR is more variable, and changes in blood pressure, reflecting changes in blood volume and cardiac output, lead to changes in GFR and urine flow, forming a simple feedback system.)

Hydrostatic pressures in glomerular capillaries are much higher than those found in other capillary beds, and in different strains of rat vary from 6 to 8 kPa (45–60 mm Hg). Pressure at all points along the capillary loops seems to be the same within about 2%, the limits of the accuracy of the experimental method (see chapter 11). Figure 9.2 represents the pressure profile of the renal vasculature of the Munich-Wistar rat, a mutant strain in which a few glomeruli are conveniently found on the surface of the kidney (most are deeper inside the cortex). In these experiments the hydrostatic pressure in the proximal tubules was 1.3 kPa (10 mm Hg), so the driving force for ultrafiltration was 6 kPa (45 mm Hg) minus 1.3 kPa (10 mm Hg) minus the colloid osmotic pressure of the plasma proteins (these are retained by the filter and so $\sigma = 1$, in contrast to smaller molecules which are freely filterable and so $\sigma = 0$). The colloid osmotic pressure was 2.7 kPa (20 mm Hg) in the afferent arteriole but, as fluid was lost by filtration and the retained proteins were concentrated, rose to equal the hydrostatic pressure difference by the time the efferent arteriole was

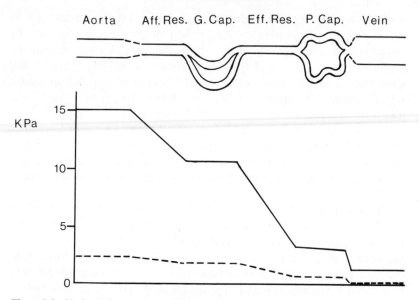

Figure 9.2 Hydrostatic pressure at different points in the circulation of the rat (solid line) and freshwater lamprey (broken line). Aff. Res. = afferent resistor (mainly afferent arterioles), G. Cap. = glomerular capillaries, Eff. Res. = efferent resistor (efferent arterioles), P. Cap. = peritubular capillaries (based mainly on data in Brenner *et al.*, 1974 and unpublished observations of McVicar and Rankin on 5 freshwater lamprey kidneys).

reached. (The pressure profile of a freshwater lamprey kidney vasculature, included for comparison, shows how great an increase in filtration pressure has occurred in the course of vertebrate evolution.) However, since the changes which do occur in GFR are thought to result from changes in filtration fraction, there is some controversy as to whether filtration equilibrium is always reached before the end of the glomerular capillaries (Arendshorst and Gottschalk, 1980).

As the plasma ultrafiltrate passes along the proximal tubule, substances required by the body (sugars, amino acids, any proteins which might have leaked through the filtering system), are actively reabsorbed. The osmolality of the fluid does not change but the tubular fluid/plasma inulin ratio (TF/P_{inulin}) increases steadily along the length of the tubule, indicating abstraction of fluid. This is another example of water movement in the apparent absence of an osmotic gradient, and once again active sodium transport is involved. The Diamond standing gradient hypothesis (chapter 5) has been invoked, since intercellular spaces are present between the proximal tubule cells, but the assumption held for many years that there is no osmotic gradient across the tubule wall has recently been questioned (Andreoli and Schafer, 1979). The tubular epithelium is so permeable to water that very small gradients—1 to 3 mOsm kg^{-1}—would be sufficient to account for the observed rate of fluid absorption. Such small differences unfortunately are comparable with the errors inherent in osmolality measurements of tubular fluid samples, so may have escaped detection. A variety of transport mechanisms found in proximal tubules could contribute to such small luminal hypotonicities—uptake of small organic molecules (in many cases, such as amino acid and sugar reabsorption, linked to the uptake of sodium by co-transport mechanisms), active sodium uptake with chloride ions following, or sodium–hydrogen ion exchange.

Acidification of proximal tubular fluid lowers the bicarbonate concentration (by converting it to carbonic acid which rapidly diffuses down the concentration gradient into the cells) and increases the chloride concentration, creating a concentration gradient for outward diffusion of chloride ions. Since the epithelium is very permeable to chloride, this process would be sufficient to account for the slight reduction in luminal osmolality (if it indeed occurs!) This is one of many examples where analytical and manipulative techniques are pushed to their limits in the study of renal physiology, and are still unable to resolve differences in interpretation.

Whatever the precise mechanisms involved, the important fact is that 65 to 75 % of the fluid filtered at the glomerulus is reabsorbed in the proximal tubules. If the single-nephron glomerular filtration rate (SNGFR) changes, the percentage proximal reabsorption remains exactly the same—a phenomenon known as glomerulotubular balance. Several explanations have been offered for this apparently simple relationship, which in fact is not, since proximal fluid reabsorption is dependent on active sodium uptake processes, which would have to be capable of sensing the rate at which sodium was passing in order to enable different volumes of fluid to be transported to keep the percentage uptake constant as SNGFR varies. Autoregulation of renal blood flow and of GFR and SNGFR and glomerulotubular balance all operate to ensure a constant flow of tubular fluid to the more distal parts of the nephron where the adjustments in volume and composition necessary for osmoregulation take place.

The medullary concentrating mechanism

The loop of Henle passes down into the medulla of the kidney, through regions of steadily increasing osmolality, finally forming a hairpin loop and returning to the cortex (figure 9.3). It was suggested in 1942 by Kuhn and Ryffel that such an arrangement could function as a countercurrent multiplier in which small differences in concentration of some substance between the two limbs of the loop could be magnified as fluid flowed round the system, until large concentration differences were built up

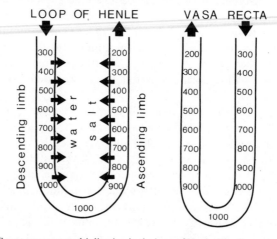

Figure 9.3 Countercurrent multiplication in the loop of Henle. The diagram shows the effect of salt movement out of the ascending limb on the osmolarity (in mOsm l^{-1}) of tubular fluid. Relatively small amounts of salt transferred from the ascending limb to the interstitial fluid will cause osmotic movement of water out of the descending limb. Small differences in concentration between adjacent points on the two limbs will lead to a large difference in concentration between the top and bottom of the loop—the *countercurrent multiplier effect*. The vasa recta blood vessels form a countercurrent system which prevents the washout of the concentration gradient established by the loops of Henle.

between the two ends of the loop. This idea was at first completely ignored, and took many years to gain acceptance, the practical problems of investigating loop functions being considerable. Micropuncture of proximal and distal tubules is possible, because some tubules lie near the surface of the kidney, and in some desert rodents the tips of the longest loops pass into the pelvis of the kidney and are reasonably accessible. The rest of the loops run through the kidney tissue surrounded by blood vessels and, considering the amount of blood flowing through the kidneys, any attempt to expose a loop surgically is an extremely hazardous procedure. Eventually, however, such an attempt was successful, and micropuncture samples taken from various points along the rat loop of Henle showed that the fluid in the descending limb was always hyper-osmotic to that in the adjacent ascending limb by just over $100\,\text{mOsm}\,\text{kg}^{-1}$ (Jamison et al., 1967).

Transfer of water from the descending to the ascending limb, or solutes from the ascending to descending limb, against relatively minor concentration gradients would lead to the establishment of a very large gradient between the top and bottom of the system. It was generally accepted that the most likely mechanism was active transport of sodium from the ascending limb into the surrounding interstitial fluid, followed either by diffusion of the sodium ions into the descending limb or osmotic abstraction of water into the interstitium, or a combination of these two processes, but all of this speculation was based on indirect evidence. It is now known that active transport of chloride ions from the thick segment of the ascending limb is the driving force for the countercurrent multiplication system.

The vasa recta blood vessels of the renal medulla also form hairpin loops. This is essential to prevent "washing out" of the concentration gradients since water and salts exchange rapidly between capillaries and the surrounding interstitial fluid. Indeed, one theoretical model of the functioning of the medulla considers the blood vessels and interstitium as a single compartment. Exactly how much water and solute passes into the vasa recta vessels and back into the general circulation obviously determines the effect the kidneys have on body fluid volume and composition, so control of the medullary circulation is of great importance, but little is known about it.

Dilute fluid enters the distal tube from the loop of Henle, and is further diluted by active uptake of salt. A number of regulatory mechanisms operate in the distal tubules and collecting ducts, the most significant from the osmoregulatory point of view being water reabsorption, sodium reabsorption and potassium secretion. The water permeability of the latter part of the distal tubule and the collecting duct is under the control of antidiuretic hormone (ADH, also called vasopressin—see below). In the presence of ADH, water moves osmotically out of the distal tubule, and the tubular fluid may become iso-osmotic with plasma before it enters the collecting duct, which passes down through the region of high osmolality in the medulla. In the absence of ADH the hypo-osmotic fluid leaving the distal tubule passes through the medulla to produce large volumes of dilute urine, since the walls of the collecting ducts are relatively impermeable to water. In the presence of ADH the permeability to water is greatly increased (figure 9.4), water passes by osmosis into the medullary interstitium (to be carried away by the vasa recta vessels), and a small volume of concentrated urine results. If the osmotic concentration in the

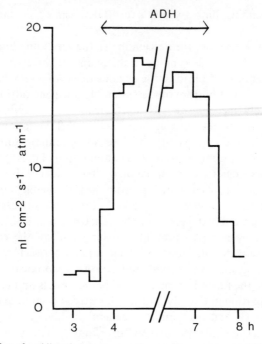

Figure 9.4 Effect of antidiuretic hormone on the osmotic water permeability of an isolated perfused rabbit collecting duct (redrawn and modified from Abramow, 1979).

collecting ducts is raised by the presence of an impermeant solute, the reabsorption of water is reduced and an *osmotic diuresis* ensues. This can be produced, for example, by administration of the non-metabolised sugar mannitol to increase urine flow in conditions of excessive water retention.

Sodium uptake and potassium secretion are under the control of the hormone aldosterone (see below) which appears to affect the two processes independently (figure 9.5). Sodium concentrations are already reduced to about half plasma levels in fluid entering the distal tubules, and can be further reduced to very low levels in sodium-depleted animals with high aldosterone secretion rates. Most of the tubular fluid potassium has been reabsorbed before the end of the loops of Henle. Potassium excretion is regulated by secretion by distal tubule cells; in most cases this can be effected by merely allowing potassium to diffuse out from the cells, where its concentration is maintained by sodium-potassium pumps on the basal cell membranes, into the lumen.

The amount of water excreted is thus controlled by ADH, and the

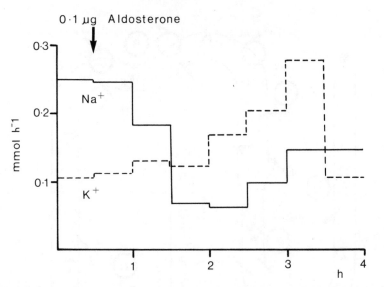

Figure 9.5 Effect of aldosterone on sodium and potassium excretion (mmol h^{-1}) in adrenalectomised rats (redrawn and modified from Morris *et al.*, 1973).

amounts of sodium and potassium ions by aldosterone, so that small or large volumes of dilute or concentrated urine can be produced as required. The main nitrogenous excretory product in mammals is urea, although some ammonia is released into the distal tubular fluid, which is soon to leave the body, thus avoiding toxicity problems. In spite of the fact that elimination of urea is one of the key functions of the kidney, this compound is not actively secreted, indeed there is considerable passive reabsorption. The rate of clearance of urea is proportional to the urine flow rate and there is a lower limit below which it cannot fall without accumulation of urea to toxic levels. This is about 500 ml per day in man, and even under conditions of extreme water deprivation, urine flow is maintained at this rate until death from dehydration ensues.

In both humans and rats, renal concentrating ability is impaired during protein deficiency, and is enhanced by dietary urea. Studies on isolated perfused tubule segments from the rabbit kidney have shown how urea contributes to the formation of concentrated urine, and this has led to a new model of the functioning of the countercurrent systems in the medulla.

Use of this technique showed that, contrary to earlier suppositions, the thin ascending limb of Henle's loop did not transport sodium. The final part of the loop, the thick ascending segment in the outer medulla (see

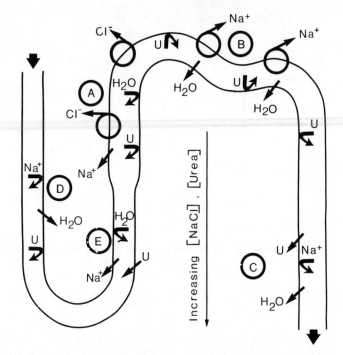

Figure 9.6 Diagram to illustrate how chloride transport from the thick ascending limb of the loop of Henle, combined with urea recycling, leads to the establishment of an osmotic gradient down the medulla. U = urea, downhill arrows represent passive diffusion, ♂ represents an ion pump, ↳ represents the inability of an ion or molecule to pass through the tubular epithelium.

figure 9.1), was found to be engaged in ion transport, but here chloride was being actively pumped out, followed by sodium. This seemed to be the active step in the creation of the osmotic gradient down the medulla. Inhibition of chloride transport by the drug furosemide reduces the osmolality of the inner medulla, and produces a large and rapidly occurring increase in urine flow. This effect is so pronounced as to be of great value in eliminating fluid which accumulates in the circulation during heart failure (when the blood pressure falls and interstitial fluid is drawn into capillaries by the colloid osmotic pressure of the plasma proteins). A great deal of research has gone into finding diuretics to treat various conditions characterised by excessive fluid retention, and the diuretic properties of furosemide were discovered before its mechanism of action was suspected.

The mechanism by which chloride secretion in the outer medulla is thought to produce high osmolalities in the inner medulla (Kokko and Rector, 1972) is complicated and will be described as a series of events referring to figure 9.6. It must of course be realised that these are not sequential steps but all proceed simultaneously; urine concentration is a continuous-flow process. The mechanisms involved are dependent on the different tubule segments having different permeabilities to water, salts and urea. These permeabilities have been determined by microperfusion studies. Since the whole process is driven by the active extrusion of chloride ions from the thick ascending limb of the loop of Henle, this forms the logical starting point for the following summary of the different stages of the process.

A. Thick ascending limb of loop of Henle (impermeable to water and urea) Active chloride extrusion, with sodium following passively, reduces the luminal osmolality.

B. Distal tubule and early collecting duct (permeable to water, impermeable to salts and urea) The tubule receives fluid of low salt concentration and osmolality, which is further reduced by active salt extrusion. The permeability of the tubules to water is greatly increased in the presence of ADH, allowing osmotic equilibration by water abstraction, and this increases the luminal urea concentration. This process continues in the first part of the collecting duct and high urea levels are attained.

C. Late collecting duct (permeable to water and urea, especially in the presence of ADH; impermeable to salts) Urea diffuses down its concentration gradient into the medullary inter- stitium, contributing to the high solute concentration. Water passes down its concentration gradient as described above.

D. Descending limb of loop of Henle (permeable to water, impermeable to salts and urea) Water passes out into the hyper-osmotic interstitial fluid, resulting in steadily increasing solute concentrations.

E. Ascending limb of loop of Henle (impermeable to water, permeable to salts and urea) Sodium and chloride concentrations have been steadily increasing during the passage of the tubular fluid along the descending limb. They now exceed the interstitial fluid concentration, so outward diffusion leads to lower luminal concentrations than in adjacent descending limbs. Urea has already been concentrated in tubular fluid passing down the descending limb but the concentration in the interstitial fluid is even higher, as a result of diffusion out of the collecting ducts. Urea therefore diffuses down its concentration gradient into the lumen of the ascending limbs, to pass on to the distal tubules, where it is further concentrated, and to the collecting ducts, where it eventually passes out into the interstitial fluid and back into the ascending limbs to complete the cycle.

The movement of sodium and chloride ions out of the ascending limb into the interstitium, to contribute to the high osmolality of the inner medulla, is thus a passive consequence of the active chloride transport from the thick ascending limb, the effect being transmitted from the outer to the inner medulla by the urea recycling process, and is not due to the presence of sodium pumps in the thin ascending limb.

Mammals thus appear to solve the problem of eliminating nitrogenous waste, whilst conserving water, in a rather complicated way. Urea is continuously lost in urine leaving the collecting ducts, but much of it is recycled before leaving the kidney to enable concentrated urine to be produced. The more urea there is to eliminate the more efficient the urine concentrating mechanism becomes. It is very dangerous to generalise, however, as most micropuncture studies are performed on rat kidneys and most microperfusion experiments on rabbit tubules, and urea may not have the same role in all species. In the sand rat, *Psammomys*, for example, urine of $6000\,mOsm\,kg^{-1}$ may be produced, but urea loading does not increase urine osmolality, whilst sodium chloride loading does. (*Psammomys* is not a true desert animal, living as it does on the fringe of the Sahara, but it feeds on halophytic plants and faces the problem of eliminating excess sodium.) The increase in solute concentration in fluid passing down the descending limb of the loop of Henle seems to be mainly due to solute addition, not water abstraction (Jamison *et al.*, 1980) and this cannot be explained by the model described above. Active salt transport from the thin ascending limb has been suggested in the hamster kidney, and a recent theoretical model (Moore and Marsh, 1980) suggests that sodium transport by the thin ascending limb is necessary to account for the concentrating effect in the inner medulla of the rat kidney. Even the rabbit kidney, on which much of the evidence for the urea recycling model was based, has recently been found to exhibit no clearly defined elevation of urine osmolality when urinary concentrations are greatly elevated

(Gunther and Rabinowitz, 1980). It still seems uncertain whether the sodium efflux from the thin ascending limb is the result of an active or a passive process.

From the point of view of comparative physiology, it is a pity that renal physiologists have concentrated almost all their efforts on animals whose kidneys functionally resemble their own rather than on these species with the most versatile osmoregulatory capabilities.

Control of renal function

1. *Antidiuretic hormone* (ADH)

Neurohypophysial hormones are probably involved in the control of renal function in all the vertebrates. In mammals their action is to decrease urine flow by increasing the osmotic permeability of the distal tubule and collecting duct wall, thus increasing water reabsorption as described above. In other vertebrates they reduce glomerular filtration rates as well

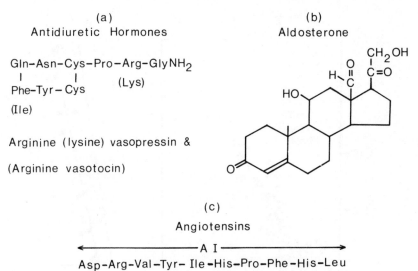

Figure 9.7 Formulae of some of the principal hormones involved in vertebrate osmo-regulation. (a) Antidiuretic hormones. The formula of arginine vasopressin is given; in lysine vasopressin arginine is replaced by lysine and in arginine vasotocin phenylalanine is replaced by isoleucine. (b) Aldosterone. (c) Angiotensins. Two amino acids are removed from AI to make AII and a further one to make AIII.

as, or instead of, increasing tubular water reabsorption. In mammals the antidiuretic hormone is arginine vasopressin (or in the case of the pig family, lysine vasopressin). In all other vertebrates it is arginine vasotocin (AVT—figure 9.7).

Classic studies on the control of ADH release were carried out by Verney (1947) who showed that, in the dog, hypertonic saline infusions causing a 2% increase in the osmolality of blood in the carotid artery caused a 90% reduction in the urine flow rate. The osmoreceptors appear to be very close to the neurosecretory cells (in the supraoptic and paraventricular nuclei) elaborating the hormone, which then passes down their axons to the posterior pituitary for storage until it is released into the blood. ADH is released only when plasma osmolality exceeds a certain threshold (figure 9.8), but this threshold is affected by blood volume changes detected by baroreceptors in the left atrium. Changing from a lying to a standing position slightly reduces the return of blood to the heart and reduces the ADH release threshold from 281 to 278 mOsm kg^{-1},

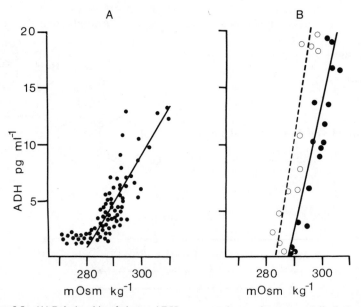

Figure 9.8 (A) Relationship of plasma ADH concentration to plasma osmalality in healthy adults. (B) Relationship of plasma ADH concentration to plasma osmolality in control rats (closed circles) and in rats given an intraperitoneal injection of polyethylene glycol to reduce their blood volume by about 15% (open circles) (redrawn and modified from Robertson *et al.*, 1977).

and infusion of hypertonic saline to increase blood volume increases the threshold to $282 \, mOsm \, kg^{-1}$. These changes are statistically significant, but a serious drawback to research in this area is that the osmoreceptors can obviously measure plasma osmolality more accurately than can laboratory osmometers!

2. The renin–angiotensin–aldosterone system

Each distal tubule of the mammalian kidney comes back to its own glomerulus and passes between the afferent and efferent arteriole, forming a structure known as the juxtaglomerular apparatus, consisting of a region of modified distal tubule cells (the macula densa), modified afferent arteriole cells (the juxtaglomerular cells), and a mass of cells resembling the mesangial connective tissue cells of the glomerulus, the polkissen (Barajas, 1979). The juxtaglomerular cells produce a hormone, renin. This is an enzyme which cleaves a plasma gamma globulin called angiotensinogen (or angiotensin substrate) to produce a decapeptide, angiotensin I. This is degraded by a converting enzyme present in the plasma to form the octapeptide angiotensin II which is a potent vasoconstrictor. Angiotensin II acts on the zona glomerulosa of the adrenal cortex to stimulate the secretion of the mineralocorticoid aldosterone, but it may first have one amino acid removed to convert it to angiotensin III.

The first property of this system to be investigated was its role in controlling blood pressure. If blood flow in a renal artery is restricted, either experimentally or by disease, large quantities of renin are released, and the angiotensin formed causes general vasoconstriction which produces a considerable increase in systemic blood pressure, often necessitating the surgical removal of a diseased kidney. The increased pressure tends to restore an adequate blood supply to the kidney, but this mechanism is probably of importance only in pathological conditions.

The concentrations of angiotensin needed to produce smooth muscle contraction are greater than those which stimulate aldosterone release (figure 9.9). Angiotensin II is a more potent vasoconstrictor than angiotensin III but the latter is slightly more potent in releasing aldosterone. However, recent evidence shows that, in the sheep, the rate of aldosterone release is not proportional to plasma angiotensin II and III levels thus suggesting that other factors are involved (Blair-West et al., 1979). No aspect of renal physiology ever seems to be simple or straightforward! Angiotensin II can also act on the brain to stimulate drinking (see chapter 7).

Plasma renin, angiotensin and aldosterone levels are elevated during

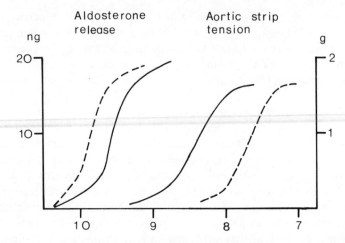

Figure 9.9 Dose response curve for the action of angiotensin II (solid line) and angiotensin III (broken line) on aldosterone release from rabbit adrenal cell suspension (ng 10^5 cells^{-1} h^{-1}) and on rabbit aortic strip tension (redrawn and modified from Davis and Freeman, 1976).

sodium depletion (Table 9.1) but the precise stimulus for renin release is uncertain. Reduced plasma sodium concentration will, with a constant GFR, lead to less sodium being filtered and, with a constant proportion reabsorbed by the proximal tubule, will lead to a reduced distal delivery of sodium. Confusing results have been obtained in experiments where the sodium concentrations in fluid perfusing the distal tubule were varied. Recent suggestions about the factor determining the rate of renin release include the rate of uptake of chloride ions by the macula densa cells, or that it is the rate of delivery of ions to the distal tubule, rather than their concentration in the tubular fluid, which is important (Churchill *et al.*, 1978). The aldosterone produced stimulates sodium reabsorption in the distal tubules, which leads to reduced sodium excretion and the restoration of plasma levels, provided a minimal salt intake is maintained.

3. *Autoregulation of renal blood flow and GFR*
The kidney has several finely-balanced mechanisms to ensure that constant amounts of fluids and salts—a small proportion of the filtered loads— are delivered to the distal tubules. From then on the water and salt regulatory mechanisms (ADH; aldosterone plus other so far unidentified factors) operate to adjust the amounts excreted to suit the body's needs. Flow through the distal tubule can be artificially manipulated *in situ* by

Table 9.1 Comparison between Australian rabbits from a desert region where the vegetation is rich in sodium and a sodium-deficient alpine region—from Blair-West *et al.*, 1968.

	Renin (Units/g kidney cortex)	Aldosterone (ng/100 ml blood)	Urine concentrations (mmol l^{-1}) Na	K
Alpine	162	130	0.6	208
Desert	30	9	139	319

microperfusion techniques in a tubule in which a proximal oil block has been inserted by micropipette to prevent the normal flow of tubular fluid. SNGFR can be measured by collection of fluid proximal to the block, or proximal stop-flow pressure (the pressure in the tubule lumen which balances the filtration pressure in the glomerular capillaries) can be monitored using a micropipette connected to a servo-null pressure measuring device (chapter 11) whilst flow through the distal tubule can be varied from zero upwards. As distal flow is increased, SNGFR decreases showing that a feedback mechanism operates at the single nephron level. Stop-flow pressure also declines, so that pressure in the glomerular capillaries must have decreased, presumably as a result of constriction of the afferent arterioles. This will tend to return both renal blood flow and GFR to normal, following increases which lead to increased flow along the tubules.

The juxtaglomerular apparatus appears to be in the ideal position to act as a feedback link between the distal tubule and the glomerular blood supply. However, attempts to implicate this system in autoregulation at the single nephron level have led to great controversy. It has been suggested that increases in distal tubular fluid flow are detected by the macula densa cells, which stimulate local release of renin from the juxta-glomerular cells, and locally produced angiotensin II causes contraction of the afferent arteriole. However, some results (Churchill *et al.*, 1978) suggest that a *decrease* in distal tubular fluid flow caused an *increased* renin release into the general circulation (which is not to deny the somewhat improbable possibility of opposite effects on local and systemic release). Autoregulation of GFR still occurs in the presence of inhibitors and blockers of the renin angiotensin system, and changes in renal renin content do not affect autoregulatory capabilities, so the paradoxical situation of a mechanism which would appear to be ideally suited for relating distal tubular flow to SNGFR, yet is not involved in the feedback mechanism, may be the reality, although current research could rapidly outdate these remarks.

One possible reason for discrepancies between experimental results is that the response of the macula densa to a stimulus may be modulated by a receptor within the juxtaglomerular apparatus sensitive to interstitial fluid pressure. It may be that feedback activation is augmented by extracellular fluid volume depletion, possibly to the point where SNGFR can be reduced even though distal fluid delivery is normal.

4. Glomerulotubular balance

A variety of theories have been advanced to explain how the proximal tubule is able always actively to reabsorb a constant proportion of the filtrate entering it. Stretching of the tubule wall to allow greater fluid absorption has been suggested, and a great deal of work has been done on the theory that changes in colloid osmotic pressure in the peritubular capillaries influence fluid movement. Increases in GFR arising from an increase in filtration fraction will lead to a higher protein concentration in post-glomerular blood. This could form a neat self-regulatory system if the increased colloid osmotic pressure increased the rate of entry of interstitial fluid into the capillaries and this speeded up absorption across or between the proximal tubule cells, to balance the increased amount of fluid entering the tubule. Conflicting results have led many workers to abandon these theories, and an idea currently attracting attention is that an unknown substance present in the glomerular ultrafiltrate regulates proximal fluid reabsorption, stopping further reabsorption when its concentration has risen to the correct level. Support for this theory is found in experiments in which glomerulotubular balance could be demonstrated in perfused rat proximal tubules when the perfusate was rat plasma ultrafiltrate, but not when Ringer's solution was used.

5. Other hormonal regulatory mechanisms

When blood volume is expanded with no change in osmolality, either by isotonic saline infusion or infusion of blood from a reservoir which has been previously allowed to mix and equilibrate with the animal's own blood, diuresis and natriuresis (increased sodium excretion) occur. Cross-circulation experiments in which the blood of two rats is allowed to exchange show that if the blood volume of one animal is expanded and that of the other kept constant, both animals experience the "volume diuresis". This suggests that a hormone produced by the volume-expanded animal has crossed to the other animal to produce diuresis and natriuresis. This hormone has been called "sodium-excreting hormone" and has not yet been identified, although two natriuretic substances have been isolated from human urine (de Wardener, 1978).

Kallikrein, like renin, is a renal hormone which acts enzymatically on a plasma protein to produce a decapeptide, bradykinin, which in this case is a vasodilator. Both the renin–angiotensin system and the kallikrein–bradykinin systems interact with prostaglandins (fatty acid tissue hormones, with widespread effects, including vasodilation, released under a variety of conditions) which are produced in large amounts in the renal medulla. They may be involved in the control of medullary blood flow.

The osmoregulatory versatility of the mammalian kidney

The osmoregulatory capabilities of mammals are largely determined by the concentrating abilities of their kidneys. Eliminating large water loads presents no difficulty, as any serious beer drinker can demonstrate. (In fact, beer drinkers over-produce urine due to the effect of alcohol on the hypothalamic osmoreceptors; the resultant dehydration could contribute to a hangover!) All that is required is a reduction in, or cessation of, ADH secretion, allowing the large volumes of fluid delivered to the collecting ducts to pass through the medulla with little absorption. A disease in which ADH is not produced, *diabetes insipidus*, is characterised by very high rates of urine flow (more than 20 litres per day in humans). A mutant strain of rats with this disorder is used in experimental work—they produce 250 ml urine per kg body weight per day, but as long as they have plenty of water to drink can maintain normal water balance. The more common form of diabetes, *diabetes mellitus*, was so called because the urine was not only copious but tasted sweet (in the days when clinical tests were rather less sophisticated than they are now). This disease, caused by lack of insulin, results in elevated plasma glucose levels and, because the maximal uptake rate of the proximal tubular reabsorptive mechanism is exceeded, glucose passes into the collecting ducts and provokes an osmotic diuresis.

Producing concentrated urine is more of a problem, and mammals are limited in their ability to conserve water by the structure of their kidneys. The solutes to be excreted also determine the maximum urinary concentration, and therefore the minimum water loss. Since the human kidney can produce urine of up to $1400\,\mathrm{mOsm\,l^{-1}}$ it might be thought that we could drink sea water ($1000\,\mathrm{mOsm\,l^{-1}}$) in an emergency and have a useful gain of free water, after excreting the salts. As Schmidt-Nielsen has pointed out this is not the case, since the maximum urinary chloride concentration we can produce is $400\,\mathrm{mmol\,l^{-1}}$, so in order to eliminate the chloride ions in a litre of sea water ($535\,\mathrm{mmol\,l^{-1}}$) we would have to excrete 1350 ml of urine, a net water loss of 350 ml. Also, figures for maximum urinary concentrations apply only to low flow rates, and more water would be lost as a result of the diarrhoea caused by magnesium sulphate in sea water, so death from dehydration would be greatly accelerated. Drinking your own urine would be of no help at all—its volume will have been reduced to the minimum needed to prevent the accumulation of toxic urea. It has been demonstrated that survival at sea for long periods with no drinking water is possible by catching teleosts, which have already performed the osmotic work of producing dilute body fluids, squeezing them and drinking the fluid. Some knowledge of animal osmoregulation would be a great asset to anyone shipwrecked, since it is important to drink teleost and not elasmobranch body fluids!

A number of desert mammals have kidneys with remarkable concentrating powers. However, it is difficult to find a better example of the versatility of the mammalian kidney than that of the vampire bat, *Desmodus rotundus* (McFarlane and Wimsatt, 1969). When given blood to feed on, these bats consumed an amount equal to 45% of their body weight in about 20 min. Rapid urine production commenced during feeding, and reached a peak rate of 250 ml kg body weight^{-1} h^{-1} (of 475 mOsm kg^{-1}) shortly after feeding stopped. This astonishingly high rate is obviously essential to allow the bat to return to its cave after feeding, loaded down with the large volume of blood it must ingest to obtain sufficient nourishment. In fact the bats are unable to take off from a level surface after a meal, but presumably launch themselves from the back of the prey animal. As they fly back to the cave, their water balance problems change completely. They lose water very rapidly, both from the respiratory tract and from the large wing area, and as the nutrients in the blood (almost entirely protein) are digested, large amounts of urea are produced. The problem now is to eliminate the urea with minimal water loss, and 7 h after a meal they produce urine of up to 6000 mOsm kg^{-1}, with urea concentrations of up to 5000 mmol l^{-1}.

This is a neat illustration of the way kidney function can be adapted to meet changing requirements in the mammal, but unfortunately nothing is known about the regulatory mechanisms involved. There is clearly no lack of research problems for the osmoregulatory physiologist.

CHAPTER TEN

SPECIALISED OSMOTIC PROBLEMS

SO FAR IN THIS BOOK WE HAVE BEEN CONCERNED WITH THE INTERACTION
between osmotic control mechanisms and the total osmotic/ionic stresses
imposed upon animals by their environment. In this chapter examples are
given to show that ionic or osmoregulatory mechanisms may be involved
in the adaptive responses of animals to problems which affect areas of their
physiology divorced from their overall osmotic control (e.g. buoyancy in
cephalopods), or which relate to stresses which do not have obvious
osmotic effects (e.g. low temperature exposure, or acid or alkaline
environmental pH). They have been included to illustrate the importance
of considering osmoregulatory implications when investigating a variety of
aspects of animal biology.

Low temperature survival

The body fluids of most animals have theoretical freezing points ranging
from about -0.2 to $-2.0°C$ depending upon their concentration. At
middle and high latitudes, in both aquatic and terrestrial environments,
temperatures may well fall below these levels for periods which vary from a
few hours, in the case of littoral animals exposed to the influence of cold air
by the falling tide, to many months in the case of organisms living in Arctic
terrestrial or freshwater habitats during winter. Some of the mechanisms
which are employed by ectothermic animals to survive low temperatures
(e.g. supercooling and high tissue glycerol levels in insects, antifreeze
substances in the blood of Antarctic ice fish) fall outside the scope of this
book, but in most cases there is an osmotic involvement. Two categories of
animals may be considered, those which tolerate freezing of the body fluids
and those which avoid freezing.

A. Animals which freeze

Some terrestrial invertebrates, particularly insects, survive low temperatures by supercooling, and provided they are not in contact with ice crystals their body fluids do not freeze. Obviously a dry integument is necessary for such survival as any external moisture will freeze and inoculate the interior of the animal with ice crystals. However terrestrial animals with moist body surfaces, freshwater or littoral animals, and many terrestrial insects which are exposed to temperatures below the supercooling point of their body fluids, cannot avoid freezing.

From all of the available evidence it is apparent that intracellular freezing is always lethal; frost-resistant animals can tolerate freezing of the extracellular fluids only. However, it is becoming increasingly clear that death at low lethal temperatures is caused not by ice finally penetrating the cells, but by osmotic damage. As ice is formed in the body fluids, the remaining unfrozen fluid becomes more concentrated until eventually a damaging osmotic pressure exists across the cell membrane and reduces the cell volume below a critical point. The lethal limit for many species of littoral molluscs appears to correspond to the temperature at which 60–70% of the body water is frozen. Thus in *Mytilus edulis* (which dies at −10°C) and *Venus mercenaria* (which survives to −6°C) the amount of body water frozen at the lower lethal temperature is the same (64%) in each case. However, a few animals are exceptions—the winkle *Littorina littorea* survives to −22°C with 76% of its body water frozen, while the littoral barnacle, *Balanus balanoides* dies when 40–45% of its body water is frozen at its summer lower lethal temperature (−6 to −7°C), but tolerates the freezing of more than 80% of its body water in the winter, when it survives to −18°C.

B. Animals which avoid freezing

Most of the animals studied so far in this group are teleost fish. In Arctic waters the temperatures of the sea (especially at depth) may fall to −1.7 to −1.8°C. This is slightly above the freezing point of sea water but below the freezing point of the blood of most teleosts (−0.5 to −0.8°C), so why don't Arctic fish freeze? For deepwater fishes (e.g. *Boreogadus saida*, *Liparis turneri*) the answer is that the animals have slightly higher osmolarities than do temperate water teleosts (blood F.P. −0.9 to −1.0°C) but not to an extent which significantly reduces their vulnerability to freezing. The fish are supercooled and survive simply because they never encounter ice crystals; if brought to the surface and touched with ice they freeze immediately.

The situation for surface fish is rather different. Sculpins (*Myoxocephalus scorpius*) and fjord cod (*Gadus ogas*) collected in the summer have a blood F.P. of $-0.8°C$, comparable with southern relatives. In winter however, when the sea water is about $-1.7°C$ the blood freezing points for both species are about $-1.5°C$. This means that the blood concentration is almost twice as high as in summer, a considerable adjustment. The generally accepted view is that the increased plasma sodium and chloride concentrations are a result of the reduced efficiency of the fish's ion transport mechanisms at very low temperatures (Vernberg and Silverthorn, 1979). However, Baltic fish living in hypo-osmotic brackish water also increase plasma ion levels at low temperatures, suggesting the contrary, i.e. that the increase is an actively controlled regulatory process (Oikari, 1975). The fish are still slightly supercooled (by about $0.2°C$) but probably risk freezing only if in prolonged contact with ice.

Invertebrates have largely been ignored in this respect; marine invertebrates are iso-osmotic with sea water so are not vulnerable to freezing unless the sea water itself freezes, when their body fluids presumably freeze as in littoral, terrestrial and freshwater forms exposed to ice contact. Recently, however, amphipods living beneath the ice in Arctic littoral pools have been studied. In the littoral zone, the air temperature may fall well below $-10°C$ and pools freeze over. When this happens the unfrozen water beneath the ice becomes more and more concentrated as fresh water is frozen out. Salinities as high as 50–60‰ develop very quickly and the unfrozen water can fall to $-8°C$. The gammarid *Gammarus duebeni* survives under the ice because it is euryhaline enough to tolerate the salinity changes, and because its body fluid concentrations increase and thus reduce the amount of supercooling. It seems likely that several other members of the invertebrate fauna of these pools survive by similar means; on the other hand shore fish, with their relatively low blood concentrations, are noticeably absent in the winter at these latitudes.

Buoyancy

Pelagic marine animals (those which swim or float, at any depth, in the sea) expend less energy if they do not have to create lift by muscular work. Many marine animals such as mackerel, tunny, some squids (e.g. *Loligo*) and the majority of pelagic elasmobranchs, have no special buoyancy mechanisms and must always swim to avoid sinking. They are like aeroplanes whose engines must run continually if they are to remain in the sky. Others have gas spaces within the body (e.g. teleost swim bladders, *Nautilus* shells, cuttle-bone, the pneumatophores of Siphonora), or they accumulate fats or lightweight ions to reduce their overall densities to values close to, or less than, sea water (1.026 g ml^{-1}) and achieve neutral or positive buoyancy. They resemble balloons and airships which either require no propulsion at all, or rely on engines solely for horizontal motion.

The accumulation of fats and oils (e.g. in the enlarged livers of some sharks) and the mechanisms of teleost swim bladders and coelenterate gas floats have no connection with osmotic or ionic regulation, but these latter processes are initially bound up with the production of gas spaces in cephalopods and the accumulation of lightweight ions in some pelagic organisms.

For our knowledge about the physiology of cuttle-bone, *Nauplius* and *Spirula* shells, we must rely heavily on the elegant work of Prof. E. J. Denton and his co-workers, who have demonstrated a common flotation mechanism in these cephalopods. For convenience, only cuttle-bone (the buoyancy organ of the cuttlefish *Sepia officinalis*) will be discussed here. The calcareous cuttle-bone (figure 10.1) is composed of numerous

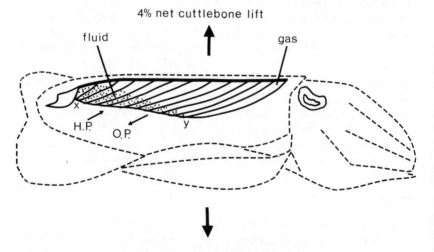

Figure 10.1 The cuttle-bone of *Sepia officinalis* (semidiagrammatic and redrawn from figures of Denton and Gilpin Brown, 1973). x–y indicates siphuncular surface, HP indicates hydrostatic pressure, OP indicates osmotic pressure.

chambers divided by septa about 0.7 mm apart; the maximum chamber volume is about 2 ml. On the ventral surface of the bone is a flat epithelium, the siphuncle, in contact with each of the chambers which are otherwise independent sealed structures. The cuttle-bone makes up about 9 % of the animal's volume and it has a specific gravity of about 0.6, so that in an animal weighing 1000 g in air it will give a lift of some 40 g, sufficient to oppose the excess weight in sea water of the rest of the animal's tissue and give near-neutral buoyancy. However, the specific gravity of the cuttle-bone is variable; in buoyant cuttlefish it is about 0.5 but in cuttlefish resting on the bottom it is nearer 0.7. Experiments have shown that cuttle-bones of S.G. 0.5 contain 10 % liquid, while those of S.G. 0.7 contain 30 % liquid, the rest of the space in the chambers being filled with gas. Clearly, varying the proportions of the liquid and gas within the bone allows the cuttlefish to control its buoyancy between slightly negative, neutral and positive.

Cuttlefish live at depths as great as 150 metres, so may be exposed to pressures of up to 15 atmospheres. Since the chambers are not hermetically sealed but are in indirect contact with the environment via the siphuncular epithelium, why does fluid not rush into the gas spaces within the cuttle-bone and destroy the buoyancy mechanism? It might be thought that the pressure of gas within the cuttle-bone would be controllable and could match the external hydrostatic pressures, but in fact the gas within the cuttle-bone varies about 0.8 atmospheres and never has a pressure greater than one atmosphere. What then opposes external hydrostatic pressure? The answer lies in the physiology of the siphuncle. This epithelium has the ability to pump salts out of the liquid within each chamber to create an osmotic pressure between the liquid and the rest of the tissues. This osmotic pressure means water tends to flow out of the chamber, and opposes the hydrostatic pressure tending to force fluid in. At the sea surface, hydrostatic pressure is negligible and the fluid within the chamber is iso-osmotic with sea water. With increasing depth, the fluid becomes more hypo-osmotic to the tissues and sea water, and the osmotic pressure across the siphuncle becomes greater. A change in amount of salt pumping will obviously result in an imbalance between osmotic and hydrostatic pressures, fluid will move into or out of the cuttle-bone chambers and the animal will rise or sink.

The theoretical limit of this mechanism will be reached when all ions are removed from the chamber liquid; since sea water and cephalopod body fluids have a concentration of about 1000 mOsmol l^{-1}, the osmotic pressure across the siphuncle would correspond to 22.4 atmospheres (see chapter 1) and could oppose a hydrostatic pressure corresponding to a depth of about 220 metres, i.e. below the depth range of *Sepia*. In fact the lower depth limit is probably set by the strength of the cuttle-bone structure itself which must resist the pressure difference between the interior and exterior of the chambers, and in pressure tests the cuttle-bone has been found to implode at a pressure corresponding to a depth of 200 metres.

The cephalopods also provide us with several examples of the light-weight ion strategy for achieving buoyancy. Gelatinous animals, with a relatively small proportion of heavy components (e.g. ctenophores, jelly-fish) achieve neutral buoyancy because the body fluids are iso-osmotic with sea water, but in which the relatively heavy divalent sulphate ion is partially replaced with chloride. In animals with a higher proportion of protein this mechanism would be ineffective, though it plays some part in the buoyancy of the gelatinous pelagic octopod *Japetella diaphana*. Far more important is the replacement of heavy cations by lightweight ammonium ions in the oceanic squids, the Cranchidae. These have a large, liquid-filled buoyancy chamber formed from the coelom (which is normally small in molluscs). The specific gravity of the fluid is about 1.01 (compared to 1.026 for sea water) despite its iso-osmocity with sea water, and is largely a solution of ammonium chloride stabilised by a low pH, which prevents ammonia diffusion out of the sac.

A final example of the interrelationship between buoyancy and osmotic/ionic regulation occurs in pelagic fish eggs. The initial osmolarity of teleost eggs when released from the female is determined by the osmoconcentration of the ovarian fluid, which is close to iso-osmocity with the blood (c. 400 mOsmol l^{-1}). In fish with demersal eggs which fall and become attached to the substrate (e.g. herring, *Clupea harengus*) the ovoplasm rapidly becomes iso-osmotic with the medium, but in fish like the cod, *Gadus morhua* or the plaice, *Pleuronectes platessa* the osmolarity remains the same or rises only slightly. The importance of this hypo-osmocity to buoyancy is probably variable because many fish eggs possess yolk which is rich in low-density fat. However, in some species such as *P. platessa*, the yolk contains no oil droplets and the low-density hypo-osmotic yolk is crucial to positive buoyancy; unfertilised eggs lose the ability to remain hypo-osmotic after several hours and consequently sink.

Life in acid and alkaline waters

A study of the osmoregulatory abilities of animals found in naturally occurring acid and alkaline waters might lead to the conclusion that the effects of these conditions were minimal. For example, large parts of the Amazon basin have waters of pH 4–5 and an enormous variety of aquatic life is present, and apparently thriving in spite of the additional handicap of extremely low salt concentrations (sodium, calcium and chloride concentrations often being less than $0.02 \, \text{mmol} \, l^{-1}$). Some organisms can survive in alkaline lakes containing high concentrations of carbonate and bicarbonate ions and with a pH of 10 or more, although here species diversity is drastically reduced. One way of illustrating the real nature of the limitations placed on osmoregulatory processes by these extreme conditions is to study the responses of species which have not evolved the necessary adaptations.

Acid waters
Studies on the effects of water acidification on trout have been carried out because of concern about the effects of pollution. Brook trout kept at pH 3.5 show an increased sodium efflux and a reduced sodium influx, and after a few hours lose half the body sodium and die. Survival time can be accurately predicted from the rate of net loss of sodium. Increasing the

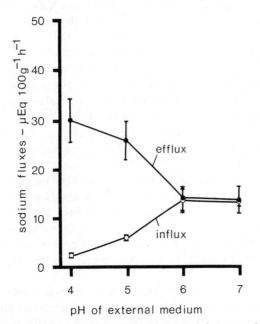

Figure 10.2 Sodium influxes and effluxes in brown trout, *Salmo trutta*, kept in waters of differing acidity (redrawn from McWilliams and Potts, 1978).

external sodium chloride concentrations to $150\,\text{mmol}\,l^{-1}$ slows the changes and prolongs survival (Packer and Dunson, 1970).

Studies on brown trout have shown that the gills are very permeable to protons, which therefore produce a large positive internal electrical potential, and this seems to be the main reason for the increased sodium loss. Increasing the external calcium concentration reduces the branchial permeability to both protons and sodium ions. The reduced sodium uptake at low pH is presumably a consequence of interference with the sodium/proton exchange mechanism (chapter 4). Figure 10.2 shows the total efflux and influx rates for sodium in brown trout. At pH 6 and above, influx balances efflux, but in more acid media the fish is in negative sodium balance (McWilliams and Potts, 1978). Acid-water fish must have evolved adaptations to prevent such effects but these have not been studied in detail.

Alkaline waters

Insight into the osmoregulatory problems of life in alkaline waters has been provided by the teleost fish *Sarotherodon (Tilapia) grahami* found in lakes fed by hot springs in the East African Rift Valley where the water temperatures may reach 40°C and pH 10.5. In laboratory experiments on *S. grahami*, a sudden increase in ambient pH seems to induce a decrease in branchial permeabilities to sodium and chloride ions; in *S. mossambicus* there was a large increase in the permeabilities (Maetz and de Renzis, 1978). *S. grahami* gills must be very impermeable to bicarbonate since a 10- to 15-fold concentration difference exists between plasma and lake water, but doubt has recently been expressed as to whether it does in fact osmoregulate in its natural environment (Skadhauge *et al.*, 1980).

Larvae of the mosquito *Aedes campestris* in alkaline salt lakes in North America are clearly capable of osmoregulation, and when kept in sodium bicarbonate solution of $700\,\text{mOsm}\,l^{-1}$ they maintain their haemolymph osmolarity at half this value. They drink the solution in order to obtain food and absorb all the water and ions ingested (Phillips and Bradley, 1977). The rectum secretes sodium ions, and presumably also bicarbonate ions, although bicarbonate fluxes have not yet been measured, as measurement of the small amount of bicarbonate excreted by 8 mg animals is so difficult.

It seems that a low permeability of the integument either to protons or bicarbonate ions is essential for survival in either acid or alkaline waters, unless mechanisms for active expulsion have evolved, and clearly further research into such mechanisms is required.

CHAPTER ELEVEN

TECHNIQUES IN OSMOREGULATORY RESEARCH

A COMPLETE DESCRIPTION OF ALL THE TECHNIQUES EMPLOYED BY RESEARCH workers who study the behavioural and physiological mechanisms of osmoregulation in animals would occupy several volumes and would rapidly become obsolete. This chapter aims only to give a brief outline of commonly-used methods and the sequence in which they are likely to be employed, the idea being to give a brief impression of the sorts of ways in which osmoregulatory physiologists work and to explain some of the more unusual methods mentioned in earlier chapters.

A. Physiological techniques

On encountering a species not previously studied, a worker interested in osmotic phenomena will first wish to know the concentration of the body fluids; if the animal's habitat is marine or brackish the influence of external salinity will be studied to determine whether the species is an osmo-regulator or osmoconformer. Firstly samples of body fluid must be obtained. Extracellular fluid such as blood plasma or haemolymph is relatively easy to obtain, in most cases by use of a syringe and needle. Samples of fluid from smaller compartments within the body, for example renal tubular fluid, require micropuncture techniques which will be described below. In all cases great care must be taken in handling the samples; a few examples will illustrate some of the problems which may be encountered. Vertebrate blood chloride concentrations vary with the degree of oxygenation of the blood because of the "chloride shift" between red cells and plasma; accurate values can be obtained only in blood collected and centrifuged under oil to prevent oxygen exchange with the atmosphere. Plasma potassium levels are low, so the slightest disruption of blood cells, caused by applying excessive negative pressure to the syringe or forcing the blood out too rapidly through the needle, leads to leakage and apparently high concentrations, which are sometimes misleadingly reported in the literature. Blood should be centrifuged immediately after collection; exchange of ions between cells and plasma proceeds quite rapidly. Breakdown of proteins into smaller molecules with a corresponding rise in the osmotic concentration can be caused by bacterial action in samples stored at room temperature or by alternate freezing and thawing. It is always best to analyse samples as soon as possible.

Measurement of osmotic concentration is usually carried out by measurement of the depression of the freezing point of water produced by the presence of solutes, but sometimes another colligative property (i.e. property dependent on the number of solute molecules present), such as depression of the vapour pressure of water, is used. Since most solutions supercool, their freezing points are difficult to measure, and two approaches to this problem are used. In the first small (down to less than 1 nl) sample droplets in liquid paraffin are frozen and then observed through a microscope while their temperature is slowly raised until the last crystal melts, giving the melting point (which is equal to the freezing point). The apparatus designed by Ramsay and Brown (1955) has been in use for many years giving extremely accurate results, but more convenient electronic nanoliter osmometers operating on the same principle are now available.

The other approach, used in most commercial osmometers, is to supercool samples and then induce freezing by mechanical vibration. The temperature rises to the freezing point and remains there until all the sample is frozen, the freezing point is then measured and converted directly into a readout in mOsmoles. This type of osmometer has been developed for clinical use and the minimum sample size (normally 200 μl) is rather large for many applications, although it is much more convenient to use than melting-point osmometers, which require peering down a microscope in order to decide when a crystal has melted. In recent years, the vapour-pressure osmometer has attracted some interest from zoologists (it has long been popular with botanists). With this apparatus a small volume of fluid (c. 5 μl) is taken up on a filter paper disc and placed in a closed chamber; or, in another model, drops of solution and solvent are placed on adjacent thermistor beads in a solvent-saturated atmosphere, where the difference in vapour pressure causes a difference in temperature between the two thermistors. In either case the resultant vapour pressure (lower than would be the case for pure water) is transduced into an electric potential; again the results are displayed as mOsmoles.

The next step in the analytical process is likely to be the determination of various ionic concentrations and ratios. Metallic cations such as sodium, potassium and calcium ions can be measured by flame photometry but (except in the case of sodium) more accurate measurements of these and other elements (e.g. magnesium) can be made using atomic absorption spectrophotometry. In most cases large dilutions of biological samples are required and accuracy and prevention of contamination are important since final readings are taken on samples of a few hundred μmoles l^{-1} or less. Problems of interference between different elements must often be considered; these are far greater in emission than in absorption measurements. Ammonium ions and anions are usually measured by various more or less complicated titration procedures. In some cases equipment is available to carry out titrations automatically, e.g. chloride meters which precipitate silver chloride in the samples by electrical generation of silver ions. The end point is determined by the increase in silver concentration, detected electrically, and the amount of chloride present is proportional to the amount of silver generated. Measurement of bicarbonate concentrations presents special problems. Bicarbonate differs from all the other ions present in biological systems by the transient nature of its existence, being in equilibrium with dissolved CO_2 which is continually being produced by metabolism and lost by respiration. The equilibrium depends on the pH of the solution, the partial pressure of CO_2 and the presence or absence of carbonic anhydrase. Measurement is not straightforward; it is necessary to acidify the sample and measure the amount of CO_2 evolved. For these reasons bicarbonate concentrations are often not directly measured but deduced from the balance of all the other ions which can be measured more easily.

Ion-sensitive glass or liquid ion-exchange electrodes are now increasingly applied. Their ion-selectivity characteristics resemble those of biological systems (chapter 1). It must however be remembered that the electrodes are never *completely* specific for one ionic species, although electrodes having a high degree of selectivity for all the most important anions and cations are available. A recent development is the manufacture of ion-sensitive microelectrodes to measure ionic activities in cells. These electrodes measure ionic *activities*, not concentrations; in fact this is the most important parameter in considerations of osmotic concentrations. To give an example, vertebrate plasma calcium activities determined using calcium electrodes are much lower than values obtained by atomic absorption spectrophotometry—this is because some of the calcium is bound to proteins, and bound ions obviously exert no osmotic effects.

Measurements of solute concentrations in whole tissues can be made after samples have been dissolved, usually in strong acids or alkalis. But as tissues contain both intracellular and extracellular fluid, intracellular solute concentrations cannot be directly determined. Although collection of interstitial fluid is extremely difficult, its solute concentrations can be calculated if an extracellular-fluid compartment such as blood plasma has been sampled and analysed. (In the case of ions, allowance has to be made for a Donnan distribution since proteins are retained in the plasma, but the correction is small and is often ignored.) The volumes of the extra- and intracellular fluid in the sample have then to be measured before the intracellular solutes can be calculated by subtracting the amounts present in the extracellular fluid from the total sample content. Since the total water content can be measured by evaporation to dryness, it is the proportion of the water in the extracellular compartment which must be determined.

Measurement of body fluid compartments

The volume of a fluid compartment which cannot be directly measured can be calculated using the dye dilution technique. Consider the case of an underground tank of unknown volume, X l, filled with fluid. If 100 g of a dye is added to the tank and sufficient time allowed for it to dissolve and become equally distributed throughout the fluid, and a sample of the fluid is then found to contain the dye at a concentration of 0.1 g l^{-1}, it is easy to calculate that $X = 1000$ l. Exactly the same principle applies to the measurement of a body fluid volume, such as the extracellular fluid volume. The following three conditions must be satisfied:

1. The indicator substance must be confined to the compartment in question.

2. Sufficient time must be allowed for complete mixing throughout the compartment. In the example above if a sample had been taken immediately after the dye was added and before it had had time to mix, the dye concentration measured would have been very high and the calculated volume very small. The "apparent volume of distribution" would have increased with time before finally levelling off at equilibrium value. Sampling at intervals is the only way to determine when equilibration is complete.

3. The indicator substance must be easy to measure in body fluids. In practice this means that it must be a dye or, more commonly, be radioactively labelled. A variety of substances has been used to measure each of the body fluid volumes. Figure 11.1 shows how the apparent volume of distribution of three of these changes with time following

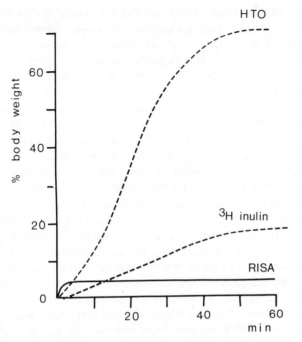

Figure 11.1   Variations in volumes of distribution with time of tritiated water (HTO), ³H inulin and radioiodinated serum albumin (RISA), which measure total body water, extracellular fluid volume and plasma volume respectively.

injection into the bloodstream of a mammal. Tritiated water, HTO, is distributed throughout the body, and is a measure of total body water. Tritiated inulin is distributed through the extracellular fluid space but does not enter cells; it therefore measures extracellular fluid volume. Radio-iodinated serum albumin (RISA) is confined to the blood system; it measures plasma volume and if haematocrit is also measured, blood volume can be calculated. One problem when making such measurements is that some of the indicator substance may be lost from the body during the equilibration period, either by renal excretion or (in the case of HTO) also by evaporation from the lungs and skin. To obtain very accurate figures the amount lost between the time of injection and the time when the equilibrium volume is calculated should be measured and only the total amount still present in the body at that time should be divided by the plasma concentration to give the volume of distribution. The interstitial and intracellular volumes can be calculated by difference if the total body water, extracellular fluid volume and plasma volume are known. In the case of tissue samples taken for analysis, the amount of tritiated inulin in the sample can be compared with the concentration in a simultaneously-taken plasma sample to determine the extracellular space in the tissue, assuming that sufficient time has elapsed since the inulin injection to allow for the equilibration.

The next steps in the analytical sequence may vary according to the interests of the experimenter. However, if a picture of body fluid osmotic pressure and ionic concentrations has already been obtained, the worker may wish to investigate osmotic processes at the cellular level. Since some of the internal osmotic pressure of cells derives from organic molecules (particularly amino acids) rather than ions, techniques have been devised to measure tissue amino acid concentrations. The first step is to measure N.P.S. (ninhydrin positive substances) in tissue samples; this is done by a colorimetric process—a fluorescent dye is produced by chemical reaction, the quantity of dye being proportional to the concentration of amino acids and ammonia in the tissue sample. The concentration of dye is estimated by measuring the fluorescence spectrophotometrically. The information gained by this process is useful but relatively crude, since it gives no indication of the relative importance of individual amino acids to cellular osmoregulation. To obtain further information the amino acid analyser must be used. This apparatus, which is very costly at present, automatically performs a complex sequence of chemical reactions interspersed with colorimetric procedures, to measure the concentrations of individual amino acids. At the time of writing, complete analysis of a sample may take as much as two days, and this, together with the use of expensive reagents and operator time, restricts the use of the apparatus for occasional osmotic studies since such equipment requires continuous use to justify the high initial cost.

Measurement of electrical potentials

Measurement of the potential difference across a cell membrane, an epithelium or the integument of an animal requires making an electrical contact with either side and connecting them via a voltmeter. For small-scale applications, contact is made by means of microelectrodes, usually filled with 3 molar KCl solution. (KCl is used because potassium and chloride ions have similar mobilities in aqueous solutions. The chloride ion mobility is about 50 % greater than that for sodium ions so if NaCl was used in the pipette, chloride ions would diffuse out of the tip faster than sodium ions, leaving it positively charged.) The micropipette is connected to a "salt bridge"—a length of tubing containing 3 molar KCl in agar. Bridges are made by filling polythene tubing with a hot agar/KCl solution which solidifies on cooling. For larger-scale applications an agar bridge can be used to make direct contact with the body fluids of an animal or the Ringer solution in an experimental chamber. The bridge is connected to KCl solution in which is immersed an electrode, usually a calomel half-cell which serves to convert the electric current from a flow of ions in the solutions or agar to a flow of electrons in the wire leading to the voltmeter avoiding errors due to junction potentials. Conduction is from the KCl solution to mercuric chloride (calomel) to mercury to the wire. Ag/AgCl electrodes can also be used. The voltmeter must be of very high input impedance to draw as little current as possible from the biological preparation. Electrode potentials and (when microelectrodes are used) tip potentials should be checked and allowed for when making readings.

Measurement of short circuit current (s.c.c.)
The s.c.c. technique was developed by Ussing for frog skin and subsequently applied to other epithelia (figure 11.2). The potential across the

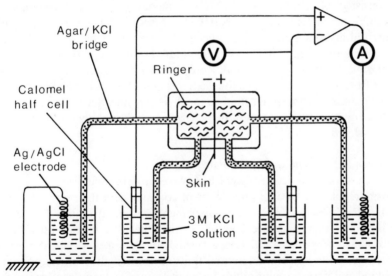

Figure 11.2 Diagrammatic representation of the apparatus used to measure short circuit current in an epithelium such as frog skin. The voltmeter (Ⓥ) records the potential difference across the skin. To short circuit the skin the positive side of the skin is connected to the inverting input (−) of an operational amplifier (▷) which produces negative feedback, passing a current which reduces the potential difference to zero. The current is measured by the ammeter (Ⓐ).

skin is measured on a voltmeter connected as described above. A current is then passed to reduce the potential to zero. This can be done manually, or automatically as shown by connecting the calomel electrodes to an operational amplifier, the positive side of the skin being connected to the reversing input. The amplifier will always pass a current in the opposite direction to that produced by the skin, thus clamping the voltage at zero. The current is passed to the skin via Ag/AgCl electrodes and agar bridges and is recorded on an ammeter. The bridges connected to the voltage detecting electrodes should be placed symmetrically as close to the skin as possible, since potential gradients will exist in the Ringer solution which has low, but not zero, resistance.

Experiments in vitro
Several different osmoregulatory organs have been studied in isolation from organisms. Many epithelia have been investigated in an Ussing-type apparatus as described above; another approach has been through perfused organs. In all these experiments provision of a suitable physiological salt solution is of great importance. Ringer's original solution consisted of 100 mequiv l^{-1} NaCl, 1 mequiv l^{-1} KCl, 1.8 mequiv l^{-1} CaCl$_2$ and 1.2 mequiv l^{-1} NaHCO$_3$. Although this was empirically arrived at from a study of the effects of various salts on heartbeat in the isolated frog heart, it closely resembles the ionic composition of frog plasma, which was unknown in Ringer's time. Very many different recipes for Ringer have since been published; it is important to use one resembling the animal's extracellular fluid as much as possible. Since pH has a large effect on transport processes the buffering system used should be carefully considered.

Dynamic studies of water and solute fluxes
The use of radioactive tracers has played a large part in the study of osmoregulatory mechanisms. We will consider here their application to the simplest case of a two-compartment system in which the rate of movement of a substance from one to the other is to be measured. Many exchanges involve several compartments, for example movement of water between cells and the interstitial fluid, interstitial fluid and blood, and blood and the external medium, but in the case of water movement in a fish only the latter step is rate-limiting so the system is considered as a simple two-compartment one. If a fish is injected with HTO and time allowed for distribution through the body water, the cumulative appearance of radioactivity in the external medium will be as shown in figure 11.3A.

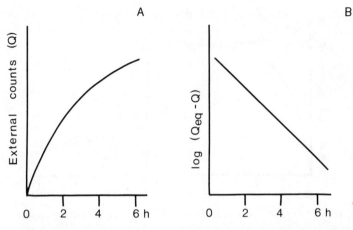

Figure 11.3 (A) Appearance of radioactivity in the external medium from a freshwater fish injected with tritiated water. (B) Plot of log (counts at equilibrium − counts at time t) against time, derived from the data in A. The turnover of tritiated water, λ, can be calculated from the slope. (Based on the results of Motais et al., 1968).

Equilibrium will eventually be reached with an amount Q_{eq} present in the tank. The quantity, Q, present at any time t will be given by the equation:

$$Q = Q_{eq}(1 - e^{-\lambda t})$$

where λ is the turnover rate between water in the fish and the external water; it equals the slope of the curve obtained by plotting $\log(Q_{eq} - Q)$ against t (figure 11.3B). The outflux of water can be calculated from the expression:

$$f_{out} = \lambda\left(\frac{V_1 \cdot V_2}{V_1 + V_2}\right)$$

where V_1 = the volume of the body water and V_2 = the volume of the external compartment.

In the case of ionic outflux from a fish the animal appears to have two compartments, one rapidly exchanging (presumably extracellular) and one slowly exchanging (presumably intracellular). Figure 11.4A shows the outflux of ^{36}Cl from a flounder in sea water; figure 11.4B shows a plot of $\log(Q_{eq} - Q)$ which is seen to consist of the sum of two exponentials. The final part of the graph is extrapolated back to zero (line x) and the values at each point in time subtracted from the original curve. This gives the line y from whose slope the fast exponential which relates to exchange between the extracellular fluid and the external medium can be calculated. In

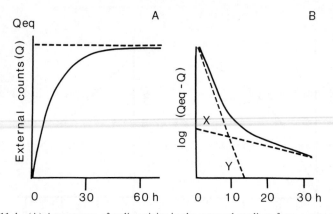

Figure 11.4 (A) Appearance of radioactivity in the external medium from a seawater fish injected with ^{36}Cl. (B) Plot of log (counts at equilibrium — counts at time t) against time, derived from the data in A. The curve is the sum of two exponentials (straight lines x and y) representing two compartments, one with a rapid and one with a slow turnover (redrawn and modified from Motais, 1967).

practice, the initial part of the curve in figure 11.4A, which is virtually a straight line, can be used as a measure of the relative outflux as described in chapter 5.

For a piece of epithelium mounted in an Ussing chamber, unidirectional fluxes are easy to measure as there is ready access to the two compartments. If fluxes in opposite directions through two adjacent pieces of epithelium of identical area are measured net flux can be calculated by difference. Under conditions where no electrochemical gradients exist (e.g. short-circuited with Ringer on both sides) this equals the rate of active transport.

Micropuncture techniques

Investigation of osmoregulatory mechanisms often involves the use of small animals, organs or parts of organs and small-scale techniques have to be devised. One of the simplest and most elegant is the method devised by Ramsay for studying secretion by insect Malpighian tubules, in which an isolated tubule is partly immersed in a small drop of solution in a bath of liquid paraffin. Fluid emerging from the end of the tubule forms another drop in the liquid paraffin and can be collected.

Micropuncture involves taking samples from anaesthetised animals with micropipettes mounted in micromanipulators. The micropipettes have first to be made by heating and drawing out glass capillary tubing to a final tip diameter of a few μm and sometimes grinding the end to make penetration easier. Evaporation is very rapid from nanolitre samples with their high surface area to volume ratio and exposure to air must be avoided. This is achieved by filling the tip of the micropipette with mineral oil such as liquid paraffin, which also covers the organ to be punctured. Once the sample has been taken, more oil from the overlying layer is aspirated to seal the pipette tip. Samples are then stored and handled under oil. Water is very slightly soluble in liquid paraffin and very small samples have been known to shrink and disappear within a few hours so it is important to use water-saturated liquid paraffin.

Having obtained a sample the next problem is how to analyse it. Osmolality measurement with a Ramsay and Brown type apparatus is well-suited to micro-samples, and micro-modifications of other methods have been developed; integrating flame photometry for example where the brief flash of light as a small sample is introduced into the flame is measured through two different filters to enable sodium and potassium concentrations to be determined. One problem is the accurate measurement of very small volumes (down to a fraction of a nanolitre), which is usually achieved by means of constriction micropipettes, made in a microforge and calibrated with highly radioactive solutions of known activity.

A very useful technique for assaying virtually every element of interest is electron probe X-ray microanalysis which can be applied to micropuncture samples by depositing them under oil on a polished surface (preferably of beryllium because of its low background), deep-freezing them, dissolving the oil away and freeze-drying at below -40°C. The samples should form uniform circular microcrystalline deposits which when bombarded with an electron beam give off X-rays of characteristic energy and wavelength for each element. Two methods of discriminating between elements are available; energy-dispersive and wavelength-dispersive analysis. The former is more convenient because all elements can be measured simultaneously but suffers from the great disadvantage that the sodium peak tends to be very small and difficult to separate from that of magnesium. This is no problem with the latter method but the number of elements which can be analysed simultaneously depends on the number of spectrometers that can be positioned around the sample chamber, usually two.

Electron probe X-ray microanalysis has also been used to study the distribution of different elements in frozen sections of organs. The subcellular distribution of various ions can be studied with much greater accuracy than is obtainable using older histochemical techniques for localisation of ions such as sodium and chloride. Dehydration of samples obviously completely disturbs the distribution of solutes and it is often forgotten that ions can diffuse at appreciable rates through ice. This does not occur at the temperatures needed to prevent the formation of large ice crystals (below − 130°C) and with the microscope stage at − 170°C, loss of water by sublimation is negligible. Section cutting is difficult at very low temperatures and for some purposes may not be necessary; since penetration of the electron beam into tissues is very poor, adequate resolution may be obtainable by looking at the planed surface of a block of tissue. Use of this technique has shown that chloride cells (chapter 5) do indeed contain a high concentration of chloride but whether this is within the cytoplasm or the tubular system is beyond the resolving power of any system currently in use.

Micro-pressure measurement using a servo-null apparatus
This technique is used in osmoregulatory research only in studies on vertebrate renal physiology. A micropipette filled with 2 molar NaCl solution is inserted into, say, a capillary containing the equivalent of 0.15 molar NaCl. An electric circuit is formed via another connection with the animal's blood system and the resistance of the circuit monitored. Almost all the electrical resistance is due to the very narrow conducting pathway through the pipette tip. If plasma enters the tip because it is at a higher pressure than the fluid in the pipette, the electrical resistance of the circuit increases due to the lower conductivity of the plasma. This causes a feedback system to operate a pump which pushes more 2 molar NaCl into the micropipette until the resistance returns to its pre-set value. The system operates to keep the plasma/NaCl interface in a constant position in the pipette tip thus ensuring that the pressure in the 2 molar NaCl solution, which is monitored by a pressure transducer, is identical to that in the plasma. Because of the fast response time of the system it is possible to observe heartbeats in a capillary—without this it would be impossible to tell if the pipette tip was correctly positioned in the lumen, especially in glomerular capillaries which are not visible through the Bowman's capsule. Although modern pressure transducers operate with a very small volume displacement, if one were to be connected directly to a fine-tipped micropipette it would take a very long time for sufficient plasma to flow through the pipette tip to produce a reading and in any case there would be a pressure gradient through the high resistance tip, making the use of an indirect method such as that described above essential.

B. Behavioural methods

Investigations into the behavioural osmotic responses of animals form a natural three-step progression. Firstly, there is the simple question of determining whether animals can or cannot detect favourable or unfavourable external osmotic conditions. Next, efforts may be made to establish critical or threshold ionic/osmotic concentrations for such responses. Finally, if responses do exist then the anatomical site(s) responsible for sensing the external environment may be located and studied.

To find out whether animals can discriminate between various osmotic/ionic levels, Y-mazes or choice boxes of a variety of designs may be used; a few examples are displayed in figure 11.5. Significant choice behaviour in response to ionic concentrations, osmotic pressures or salinities has been demonstrated in several organisms, particularly amongst fish and crustaceans. Sessile animals may simply be observed directly; the closure responses to low salinity exhibited by several bivalves and barnacles were first noted in this manner many years ago.

To determine threshold concentrations for behavioural reactions used to require many tedious experiments. For example, let us assume that we are interested in a marine animal which might be expected to avoid low salinities. If it is a sessile species such as a bivalve mollusc we may place groups of animals in 100%, 80%, 60%, 40%, 20% and 0% sea water. Already this involves the use of many animals, since a minimum of 5 or 6 must be used for each seawater concentration to obtain statistically reliable results (similar considerations would apply if the species was a mobile one tested by choice experiments). Now let us assume that the bivalves remain open and pumping in 100%, 80%, 60% and 40% sea water but keep their shell valves tightly closed in 20% and 0% sea water. This suggests that the threshold seawater concentration which triggers valve closure lies between 20% and 40% sea water. Consequently, further experiments have to be performed, perhaps at concentrations of 25%, 30% and 35% sea water. In total as many as 40–60 animals will be used in such a study and the precision of the estimate of the threshold concentration may still be rather coarse. More recently, equipment has been developed to deliver water of continually changing "quality" (whether of osmolarity or some other factor) on a reliable repeatable basis to chambers where animals may be observed. If a linear change in osmotic or ionic concentration is imposed upon 5 or 6 animals in turn, the precise concentration inducing a behavioural response can be established for each individual. By this means a good estimate of a critical concentration may be obtained with a very few animals. In sessile animals observation of behavioural responses can be facilitated by the attachment of various stress gauges and impedance pneumographs; examples are shown in figure 11.6.

Traditionally, receptor sites are located by selective ablation (i.e. removal or destruction) of structures. In a number of crustaceans (e.g.

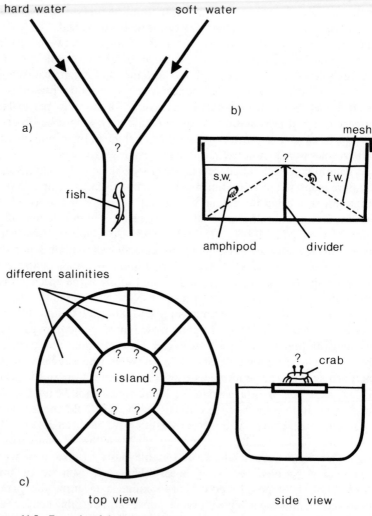

Figure 11.5 Examples of choice experiments.

crabs, amphipods) it has been established in this way that various appendages, in particular antennules, antennae and walking limbs, carry osmotic or ionic receptors. However, ablation experiments must allow for recovery from post-operative shock or haemorrhage, and for the possibility of regeneration restoring receptive capacity.

It can be important to differentiate between osmotic and ionic reception.

a) valve movement recording

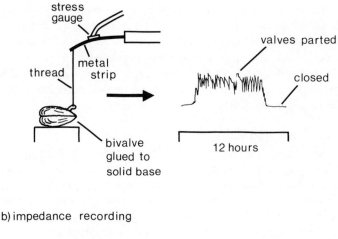

b) impedance recording

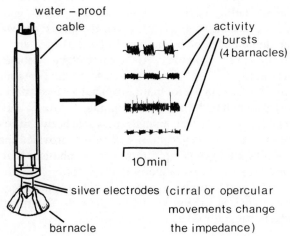

Figure 11.6 Activity-monitoring devices.

If an animal has been shown to discriminate between two seawater concentrations (say 100% and 60%) in favour of the higher concentration, it may have done so by recognising the difference in total osmotic pressure or alternatively by registering a discrepancy in concentration of a particular ion or group of ions. To clarify this, the animals can be offered a choice between 100% sea water and a solution of mannitol (or sucrose or some other non-ionic, non-toxic substance) in 60% sea water with a total osmolarity equivalence to 100% sea water. If the animals still discriminate between the two fluids then they must have an ability to detect ionic concentration differences (or perhaps react to mannitol!). On the other hand, if they do not choose between the fluids then clearly the species possesses osmoreceptors rather than structures responsive to ionic concentrations. If ionic receptors *are* strongly suggested by choice box results with mannitol solutions, then the particular ion or ions detected may be established by offering choices between solutions of similar osmolarity but different ionic composition.

Measurement of drinking rates
Drinking behaviour is obviously of great importance in the maintenance of water balance in many animals. Whilst it is a simple matter to measure the amount of water drunk by a terrestrial animal it is distinctly more difficult in an aquatic organism such as a marine teleost. The animal is placed in a tank containing a substance which is taken into the alimentary canal but not absorbed across its wall. After a suitable time interval (chosen so that a measurable quantity of the substance has entered the gut but none has reached the anus) the animal is killed, the gut excised and washed out, and the total amount of the substance present is measured and divided by the concentration in the tank to give the volume of fluid imbibed. The marker substance must be easy to measure in solutions containing gut fluids. Phenol red has been in use for many years but a wide variety of radioactive substances, including ^{51}Cr-EDTA, colloidal ^{198}Au, ^{131}I-PVP etc., have also been used. In very small animals the whole body can be radioassayed, after the external surface has been well rinsed, provided that the marker enters the alimentary canal only. It must be emphasised that drinking rate does not necessarily equal rate of water absorption into the animal, since some may be lost through the anus. Separate experiments have to be performed to measure the proportion of the water entering the alimentary canal which is actually taken up by the body.

FURTHER READING

We have not set out to provide a comprehensive bibliography of animal osmoregulation, or indeed to give a balanced survey of recent work in the field. Rather we intend to list some of the books and reviews which will fulfil these functions collectively and to give the reader the opportunity to consult the somewhat arbitrary selection of original papers cited in the text to illustrate what we consider to be some of the most interesting aspects of the subject. In some fields, notably mammalian renal physiology, progress is rapid and, although an attempt has been made to include a selection of the most recent work, no textbook can ever hope to keep up with current thinking. Fortunately, however, review articles appear frequently in these very active areas of research, and the reader is advised to consult the latest issues of journals such as *Annual Review of Physiology, Physiological Reviews* or the *American Journal of Physiology* (Section F: *Renal, Fluid and Electrolyte Physiology*), which contains regular editorial reviews, in order to bring information contained in this book up to date.

General and historical texts

Alderdice, D. F. (1972) "Responses of the marine poikilotherms to environmental factors acting in concert" in *Marine Ecology*, Vol. 1, Part 3 (ed. Kinne, O.), Wiley Interscience, London, 1659–1722.

Bentley, P. J. (1971) *Endocrines and Osmoregulation*, Springer-Verlag, New York.

Blight, J., Cloudsley-Thompson, J. L. and MacDonald, A. G. (1976) *Environmental Physiology of Animals*, Blackwell Scientific Publications, Oxford.

Florkin, M. and Schoffeniels, E. (1965) "Euryhalinity and the concept of physiological radiation" in *Studies in Comparative Biochemistry* (ed. Munday, K. A.), Pergamon Press, Oxford, 6–40.

Florkin, M. and Schoffeniels, E. (1969) *Molecular Approaches to Ecology*, Academic Press, New York.

Gilles, R. (1975) "Mechanisms of ion and osmoregulation" in *Marine Ecology*, Vol. II, Part 1 (ed. Kinne, O.), Wiley Interscience, London, 259–347.

Gilles, R. (ed.) (1979) *Mechanisms of Osmoregulation in Animals: Maintenance of Cell Volume*, John Wiley & Sons, Chichester.

Holmes, W. N. and Pearce, R. B. (1979) "Hormones and osmoregulation in the vertebrates" in *Mechanisms of Osmoregulation in Animals: Maintenance of Cell Volume* (ed. Gilles, R.), John Wiley & Sons, Chichester, 413–533.

Jørgensen, C. B. and Skadhauge, E. (eds.) (1978) *Osmotic and Volume Regulation*, Alfred Benson Symposium Series, 1977, Academic Press, New York.

Jungreis, A. M., Hodges, T. K., Kleinzeller, A. and Schultz, S. G. (1977) *Water Relations in Membrane Transport in Plants and Animals*, Academic Press, New York.

Kinne, O. (1975) *Marine Ecology*, Vol. II, Part 1, *Physiological Mechanisms* (ed. Kinne, O.), Wiley Interscience, London.

Krogh, A. (1939) *Osmotic Regulation in Aquatic Animals*, Cambridge University Press, Cambridge, reprinted by Dover Publications, New York, 1965.

Lahlou, B. (ed.) (1980) *Epithelial Transport in the Lower Vertebrates*, Cambridge University Press, Cambridge.

Lockwood, A. P. M. (1963) *Animal Body Fluids and their Regulation*, Heinemann, London.

Maloiy, G. M. O. (ed.) (1979) *Comparative Physiology of Osmoregulation in Animals*, Vols. 1 & 2, Academic Press, London.

Potts, W. T. W. and Parry, G. (1964) *Osmotic and Ionic Regulation in Animals*, Pergamon Press, Oxford.

Schmidt-Nielsen, K. (1979) *Animal Physiology: Adaptation and Environment*, 2nd Edition, Cambridge University Press, Cambridge, 560 pp.

Schoffeniels, E. (1976) Adaptations with respect to salinity. *Biochem. Soc. Symp.*, **41**, 179–204.

Schnoffeniels, E. and Gilles, R. (1970) "Osmoregulation in aquatic arthropods" in *Chemical Zoology*, Vol. V, Part A (eds. Florkin, M. and Scheer, B. T.), Academic Press, New York, 255–286.

Smith, H. W. (1953) *From Fish to Philosopher*, Little, Brown & Co., Boston (reprinted 1961, Doubleday, New York).

Wessing, A. (ed.) (1975) Excretion. *Fortschr. Zool.*, **23**, 1–362.

Chapter 1

Bonting, S. L. (1970) "Sodium-potassium activated adenosinetriphosphatase and cation transport" in *Membranes and Ion Transport* (ed. Bittar, E. E.), Wiley-Interscience, London, 257–363.

Burton, R. F. (1968) Cell potassium and the significance of osmolarity in vertebrates. *Comp. Biochem. Physiol.*, **27**, 763–773.

Diamond, J. M. and Wright, E. M. (1969) Biological membranes: the physical basis of ion and nonelectrolyte selectivity. *Ann. Rev. Physiol.*, **31**, 581–646.

Dick, D. A. T. (1979) "Structure and properties of water in the cell" in *Mechanisms of Osmoregulation in Animals: Maintenance of Cell Volume* (ed. Gilles, R.), John Wiley & Sons, Chichester, 3–45.

Edzes, H. T. and Berendsen, H. J. C. (1975) The physical state of diffusible ions in cells. *Ann. Rev. Biophys. Bioeng.*, **4**, 265–285.

Ehrenfeld, J. and Garcia-Romeu, F. (1977) Active hydrogen excretion and sodium absorption through isolated frog skin. *Am. J. Physiol.*, **233**, F46–F54.

Ehrenfeld, J. and Garcia-Romeu, F. (1980) Kinetics of ionic transport across frog skin: Two concentration-dependent processes. *J. memb. Biol.*, **56**, 139–147.

Ellory, J. C. and Lew, V. L. (eds.) (1977) *Membrane Transport in Red Blood Cells*, Academic Press, New York.

Gutknecht, J. (1970) The origin of bioelectric potentials in plant and animal cells. *Am. Zool.*, **10**, 347–354.

Harrison, R. and Lunt, G. G. (1980) *Biological Membranes. Their Structure and Function* (2nd Edn.), Blackie, Glasgow.

Helman, S. I. and Fisher, R. S. (1977) Microelectrode studies of the active Na transport pathway of frog skin. *J. gen. Physiol.*, **69**, 571–604.

Hobbs, A. S. and Albers, R. W. (1980) The structure of proteins involved in active membrane transport. *Ann. Rev. Biophys. Bioeng.*, **9**, 259–291.

Hoffmann, E. K. (1977) "Control of cell volume" in *Transport of Ions and Water in Animals* (ed. Gupta, B. L., Moreton, R. B., Oschman, J. L. and Wall, B. J.), Academic Press, London, 285–332.

House, C. R. (1974) *Water Transport in Cells and Tissues*, Edward Arnold, London.

Jhon, M. S. and Eyring, H. (1976) Liquid theory and the structure of water. *Ann. Rev. Phys. Chem.*, **27**, 45–57.

Karlish, S. J. D., Yates, D. W. and Glynn, I. M. (1978) Conformational transitions between Na^+-bound and K^+-bound forms of $(Na^+ + K^+)$-ATPase, studied with formycin nucleotides. *Biochim. Biophys. Acta*, **525**, 252–264.

Koefoed-Johnsen, V. (1979) "Control mechanisms in amphibians" in *Mechanisms of Osmoregulation in Animals: Maintenance of Cell Volume* (ed. Gilles, R.), John Wiley & Sons, Chichester, 223–272.

Koefoed-Johnson, V. and Ussing, H. H. (1958) The nature of the frog skin potential. *Acta Physiol. Scand.*, **42**, 298–308.

Lindley, B. D. (1970) Fluxes across epithelia. *Am. Zool.*, **10**, 355–364.

MacKnight, A. D. C., DiBona, D. R. and Leaf, A. (1980) Sodium transport across toad urinary bladder: A model "tight" epithelium. *Physiol. Rev.*, **60**, 615–715.

Robinson, J. D. and Flashner, M. S. (1979) The $(Na^+ + K^+)$-activated ATPase. Enzymatic and transport properties. *Biochim. Biophys. Acta*, **549**, 145–176.

Chapter 2

Bellamy, D. and Chester Jones, I. (1961) Studies on *Myxine glutinosa*—1. The chemical composition of the tissues. *Comp. Biochem. Physiol.*, **3**, 175–183.

Haldane, J. B. S. (1954) *The Origin of Life:* New Biology, **16**, 12, Penguin, London.

Harvey, H. W. (1955) *The Chemistry and Fertility of Seawaters*, Cambridge University Press, Cambridge.

Oparin, A. I. (1957) *The Origin of Life on Earth*, Oliver & Boyd, London.

Oparin, A. I. (1964) *Life, its Nature, Origin and Development*, Academic Press, New York.

Riegel, J. A. (1978) Factors affecting glomerular functions in the Pacific hagfish *Eptatretus stouti* (Lockington). *J. exp. Biol.*, **73**, 261–277.

Robertson, J. D. (1953) Further studies on ionic regulation in marine invertebrates. *J. exp. Biol.*, **26**, 182–200.

Robertson, J. D. (1957) "Osmotic and ionic regulation in aquatic invertebrates" in *Recent Advances in Invertebrate Physiology* (ed. Scheer, B. T.), University of Oregon, Eugene, 229–246.

Stolte, H. and Schmidt-Nielsen, B. (1978) "Comparative aspects of fluid and electrolyte regulation by the cyclostome, elasmobranch and lizard kidney" in *Osmotic and Volume Regulation*, Alfred Benson Symposium Series, 1977 (eds. Jørgensen, C. B. and Skadhauge, E.), Academic Press, New York.

Chapter 3

Ahokas, R. A. and Sorg, G. (1977) The effect of salinity and temperature on intracellular osmoregulation and muscle free amino acids in *Fundulus diaphanus*. *Comp. Biochem. Physiol.*, **56A**, 101–105.

Avens, A. C. and Sleigh, M. A. (1965) Osmotic balance in gastropod molluscs. 1. Some marine and littoral gastropods. *Comp. Biochem. Physiol.*, **16**, 121–141.

Bricteux-Gregoire, S., Duchâteau-Bosson, G. L., Jeuniaux, C. L. and Florkin, M. (1962) Constituents osmotiquements actifs des muscles du Crabe chinois *Eriocheir sinensis*, adapté à l'eau douce ou à l'eau de mer. *Archs. int. Physiol. Biochim.*, **70**, 273–286.

Cawthorne, D. F. (1979) *Some effects of fluctuating temperature and salinity upon cirripedes*, Ph.D. Thesis: University of Wales.

Davenport, J. (1972a) Salinity tolerance and preference in the porcelain crabs *Porcellana platycheles* and *Porcellana longicornis*. *Mar. Behav. Physiol.*, **1**, 123–138.

Davenport, J. (1972b) Volume changes shown by some littoral anomuran crustacea. *J. Mar. biol. Ass. U.K.*, **52**, 863–877.

Davenport, J. (1972c) Effects of size upon salinity tolerance and volume regulation in the hermit crab *Pagurus bernhardus*. *Mar. Biol.*, **17**, 222–227.

Davenport, J. (1972d) Study of the importance of the soft abdomen of the hermit crab *Pagurus bernhardus* in minimising the mechanical effects of osmotic uptake of water. *Mar. Biol.*, **17**, 304–307.

Davenport, J. (1979a) The isolation response of mussels (*Mytilus edulis* L.) exposed to falling sea-water concentrations. *J. Mar. biol. Ass. U.K.*, **59**, 123–132.

Davenport, J. (1979b) Is *Mytilus edulis* a short term osmoregulator? *Comp. Biochem. Physiol.*, **64A**, 91–95.

Davenport, J., Gruffydd, Ll. D. and Beaumont, A. R. (1975) An apparatus to supply water of fluctuating salinity and its use in a study of the salinity tolerances of larvae of the scallop *Pecten maximus* L. *J. mar. biol. Ass. U.K.*, **55**, 391–409.

Kinne, O. (1964) The effects of temperature and salinity on marine and brackish water animals. II. Salinity and temperature-salinity combinations. *Oceanogr. Mar. Biol. Ann. Rev.*, **2**, 281–339.

Kinne, O. (1970) in *Marine Ecology*, Vol. 1, Part 1 (ed. Kinne, O.), Wiley Interscience, New York, 407–514.

Kinne, O. (1971) in *Marine Ecology*, Vol. 1, Part 2 (ed. Kinne, O.), Wiley Interscience, New York, 821–995.

Lange, R. (1963) The osmotic function of amino acids and taurine in the mussel, *Mytilus edulis*. *Comp. Biochem. Physiol.*, **10**, 173–179.

Lange, R. (1970) Isosmotic intracellular regulation and euryhalinity in marine bivalves. *J. exp. mar. Biol. Ecol.*, **5**, 170–179.

Lange, R. and Mostad, A. (1967) Cell volume regulation in osmotically adjusting marine animals. *J. exp. mar. Biol. Ecol.*, **1**, 209–219.

Livingstone, D. R., Widdows, J. and Fieth, P. (1979) Aspects of nitrogen metabolism of the common mussel *Mytilus edulis*: adaptation to abrupt and fluctuating changes in salinity. *Mar. Biol.*, **53**, 41–55.

Lockwood, A. P. M. (1962) The osmoregulation of Crustacea. *Biol. Rev.*, **37**, 257–305.

Norfolk, J. R. W. (1978) Internal volume and pressure regulation in *Carcinus maenas*. *J. exp. Biol.*, **74**, 123–132.

Pierce, S. K. Jr. (1970) The water balance of *Modiolus* (Mollusca: Bivalvia: Mytilidae): osmotic concentrations in changing salinities. *Comp. Biochem. Physiol.*, **36**, 521–533.

Pierce, S. K. Jr. (1971) Volume regulation and valve movements by marine mussels. *Comp. Biochem. Physiol.*, **39A**, 103–117.

Pierce, S. K. and Greenberg, M. J. (1972) The nature of cellular volume regulation in marine bivalves. *J. exp. Biol.*, **57**, 681–692.

Pierce, S. K. and Greenberg, M. J. (1973) The initiation and control of free amino acid regulation of cell volume in salinity stressed marine bivalves. *J. exp. Biol.*, **59**, 435–446.

Robertson, J. D. (1960) "Osmotic and ionic regulation" in *The Physiology of Crustacea*, Vol. 1 (ed. Waterman, T. H.), Academic Press, New York, 317–340.

Schoffeniels, E. (1976) "Biochemical approaches to osmoregulatory processes in Crustacea" in *Perspectives in Experimental Biology*, Vol. 1 Zoology (ed. Spencer Davies, P.), Pergamon Press, Oxford, 107–124.

Schoffeniels, E. and Gilles, R. (1972) "Ion regulation and osmoregulation in Mollusca" in *Chemical Zoology*, Vol. VII (eds. Florkin, M. and Scheer, B. T.), Academic Press, New York, 393–420.

Shaw, J. (1961) Studies on ionic regulation in *Carcinus maenas* (L.). *J. exp. Biol.*, **38**, 135–152.

Shumway, S. E. (1977a) Effect of salinity fluctuation on the osmotic pressure and Na^+, Ca^{2+} and Mg^{2+} ion concentrations in the hemolymph of bivalve molluscs. *Mar. Biol.*, **41**, 153–177.

Shumway, S. E. (1977b) The effect of fluctuating salinity on the tissue water content of eight species of bivalve molluscs. *J. comp. Physiol.*, **116**, 269–285.

Shumway, S. E. (1977c) The effects of fluctuating salinities on four species of asteroid echinoderms. *Comp. Biochem. Physiol.*, **58A**, 177–179.

Shumway, S. E. and Davenport, J. (1977) Some aspects of the physiology of *Arenicola marina* (Polychaeta) exposed to fluctuating salinities. *J. mar. Biol. Ass. U.K.*, **57**, 907–924.

Shumway, S. E., Gabbott, P. A. and Youngson, A. (1977) The effect of fluctuating salinity on the concentrations of free amino acids and ninhydrin positive substances in the adductor muscles of eight species of bivalve molluscs. *J. exp. mar. Biol. Ecol.*, **29**, 131–150.

Shumway, S. E. and Youngson, A. (1979) The effects of fluctuating salinity on the physiology of *Modiolus demissus* (Dillwyn). *J. exp. mar. Biol. Ecol.*, **40**, 167–181.

Spaargaren, D. H. (1974) A study on the adaptation of marine organisms to changing salinities with special reference to the shore crab *Carcinus maenas* (L.). *Comp. Biochem. Physiol.*, **47A**, 499–512.

Spaargaren, D. H. (1978) A comparison of the blood osmotic composition of various marine and brackish water animals. *Comp. Biochem. Physiol.*, **60A**, 327–333.

Spaargaren, D. H. (1979) "Marine and brackish water animals" in *Comparative Physiology of Osmoregulation in Animals*, Vol. 1 (ed. Maloiy, G. M. O.), Academic Press, London, 84–116.

Stickle, W. B. and Ahokas, R. (1974) The effects of tidal fluctuation of salinity on the perivisceral fluid composition of several echinoderms. *Comp. Biochem. Physiol.*, **47A**, 469–476.

Stickle, W. B. and Ahokas, R. (1975) The effects of tidal fluctuation of salinity on the hemolymph composition of several molluscs. *Comp. Biochem. Physiol.*, **50A**, 291–296.

Thuet, P., Motais, R. and Maetz, J. (1968) Les mécanismes de l'euryhalinité chez le crustacé des salines *Artemia salina* L. *Comp. Biochem. Physiol.*, **26**, 793–818.

Tucker, L. E. (1970) Effects of external salinity on *Scutus breviculus* (Gastropoda, Prosobranchia)—I. Body weight and blood composition. *Comp. Biochem. Physiol.*, **36**, 301–319.

Chapter 4

Binyon, J. (1979) *Branchiostoma lanceolatum*—A freshwater reject? *J. mar. biol. Ass. U.K.*, **59**, 61–67.

Fettiplace, R. and Haydon, D. A. (1980) Water permeability of lipid membranes. *Physiol. Rev.*, **60**, 510–550.

Greenway, P. (1979) "Fresh water invertebrates" in *Comparative Physiology of Osmoregulation in Animals*, Vol. 1 (ed. Maloiy, G. M. O.), Academic Press, London, 117–173.

Hays, R. M., Franki, N. and Soberman, R. (1971) Activation energy for water diffusion across the toad bladder: Evidence against the pore enlargement hypothesis. *J. clin. Invest.*, **50**, 1016–1018.

Hofmann, E. and Butler, D. G. (1979) The effect of increased metabolic rate on renal function in the rainbow trout, *Salmo gairdneri*. *J. exp. Biol.*, **82**, 11–23.

Isaia, J., Girard, J.-P. and Payan, P. (1978) Kinetic study of gill epithelium permeability to water diffusion in the freshwater trout, *Salmo gairdneri*: Effect of adrenaline. *J. memb. Biol.*, **41**, 337–347.

Isaia, J. and Masoni, A. (1976) The effects of calcium and magnesium on water and ionic permeabilities in the sea water adapted eel, *Anguilla anguilla* L. *J. comp. Physiol.*, **109**, 221–223.

Keynes, R. D. (1975) "The energy costs of active transport" in *Comparative Physiology— Functional Aspects of Structural Materials* (eds. Bolis, L., Maddrell, S. H. P. and Schmidt-Nielsen, K.), North Holland, Amsterdam.

Kirschner, L. B. (1979) "Control mechanisms in crustaceans and fishes" in *Mechanisms of Osmoregulation in Animals: Maintenance of Cell Volume* (ed. Gilles, R.), John Wiley & Sons, Chichester, 157–222.

Logan, A. G., Moriarty, R. J., Morris, R. and Rankin, J. C. (1980*a*) The anatomy and blood system of the kidney in the river lamprey, *Lampetra fluviatilis*. *Anat. Embryol.*, **158**, 245–252.

Logan, A. G., Moriarty, R. J. and Rankin, J. C. (1980*b*) A micropuncture study of kidney function in the river lamprey, *Lampetra fluviatilis*, adapted to fresh water. *J. exp. Biol.*, **85**, 137–147.

Maetz, J. (1968) "Salt and water metabolism" in *Perspectives in Endocrinology. Hormones in the Lives of Lower Vertebrates* (eds. Barrington, E. J. W. and Jørgensen, C. B.), Academic Press, London and New York.

Maetz, J., Payan, P. and de Renzis, G. (1976) "Controversial aspects of ionic uptake in freshwater animals" in *Perspectives in Experimental Biology*, Vol. I, Zoology (ed. Spencer Davies, P.), Pergamon Press, Oxford.

Mangum, C. P., Haswell, M. S., Johansen, K. and Towle, D. W. (1978) Inorganic ions and pH in the body fluids of Amazon animals. *Can. J. Zool.*, **56**, 907–916.

Moriarty, R. J. (1977) *Renal function in the euryhaline lamprey (Lampetra fluviatilis* L.), Ph.D. Thesis, University of Wales.

Moriarty, R. J., Logan, A. G. and Rankin, J. C. (1978) Measurement of single nephron filtration rate in the kidney of the river lamprey, *Lampetra fluviatilis* L. *J. exp. Biol.*, **77**, 57–69.

Motais, R., Isaia, J., Rankin, J. C. and Maetz, J. (1969) Adaptive changes in the water permeability of the teleostean gill epithelium in relation to external salinity. *J. exp. Biol.*, **51**, 529–546.

Patterson, D. J. (1980) Contractile vacuoles and associated structures: their organization and function. *Biol. Rev.*, **55**, 1–46.

Payan, P. (1978) A study of the Na^+/NH_4^+ exchange across the gill of the perfused head of the trout (*Salmo gairdneri*). *J. comp. Physiol.*, **124**, 181–188.

Pedley, T. J. and Fischbarg, J. (1980) Unstirred layer effects on osmotic water flow across gallbladder epithelium. *J. memb. Biol.*, **54**, 89–102.

Prusch, R. D. (1977) "Protozoan osmotic and ionic regulation" in *Transport of Ions and Water in Animals* (eds. Gupta, B. L., Moreton, R. B., Oschman, J. L. and Wall, B. J.), Academic Press, London.

Prusch, R. D., Benos, D. J. and Ritter, M. (1976) Osmoregulatory control mechanisms in freshwater coelenterates. *Comp. Biochem. Physiol.*, **53A**, 161–164.

Riegel, J. A. (1972) *Comparative Physiology of Renal Excretion*, Oliver & Boyd, Edinburgh.

Riegel, J. A. (1977) "Fluid movement through the crayfish antennal gland" in *Transport of Ions and Water in Animals* (eds. Gupta, B. L., Moreton, R. B., Oschman, J. L. and Wall, B. J.), Academic Press, London, 613–631.

Shaw, J. (1959) The absorption of sodium ions by the crayfish, *Astacus pallipes* Lereboullet. I. The effect of external and internal sodium concentrations. *J. exp. Biol.*, **36**, 126–144.

Shaw, J. (1960) The absorption of sodium ions by the crayfish, *Astacus pallipes* Lereboullet. II. The effect of the external anion. *J. exp. Biol.*, **37**, 534–547.

Treherne, J. E. (1980) Neuronal adaptations to osmotic and ionic stress. *Comp. Biochem. Physiol.*, **67B**, 455–463.

Wendelaar Bonga, S. E. and Van der Meij, J. C. A. (1980) The role of environmental calcium in the control of prolactin secretion in the teleost *Satherodon mossambicus* (*Tilapia mossambica*). *Gen. comp. Endocr.*, **40**, 342.

Wetzel, R. G. (1975) *Limnology*, W. B. Saunders, Philadelphia.

Wright, E. M. (1977) "Passive water transport across epithelia" in *Water Relations in Membrane Transport in Plants and Animals* (eds. Jungreis, A. M., Hodges, T. K., Kleinzeller, A. and Schultz, S. G.), Academic Press, New York, 199–213.

Chapter 5

Babiker, M. M. and Rankin, J. C. (1978) Neurohypophysial hormonal control of kidney function in the European eel (*Anguilla anguilla* L.) adapted to sea-water or freshwater. *J. Endocr.*, **76**, 347–358.

Babiker, M. M. and Rankin, J. C. (1979) Factors regulating the functioning of the *in vitro* perfused aglomerular kidney of the angler fish, *Lophius piscatorius* L. *Comp. Biochem. Physiol.*, **62A**, 989–993.

Chan. D. K. O., Phillips, J. G. and Chester Jones, I. (1967) Studies on electrolyte changes in the lip-shark, *Hemiscyllium plagiosum* (Bennett), with a special reference to hormonal influence on the rectal gland. *Comp. Biochem. Physiol.*, **23**, 185–198.

Chester Jones, I., Chan, D. K. O. and Rankin, J. C. (1969) Renal function in the European eel (*Anguilla anguilla* L.): Changes in blood pressure and renal function of the freshwater eel transferred to sea water. *J. Endocr.*, **43**, 9–19.

Conte, F. P. (organiser) (1980) "Biology of the chloride cell," Jean Maetz memorial symposium. *Am. J. Physiol.*, **238**, R139–R276.

Degnan, K. G., Karnaky, K. J. and Zadunaisky, J. A. (1977) Active chloride transport in the *in vitro* opercular skin of a teleost (*Fundulus heteroclitus*), a gill-like epithelium rich in chloride cells. *J. Physiol.*, **277**, 155–191.

Degnan, K. G. and Zadunaisky, J. A. (1980) Passive sodium movements across the opercular epithelium: The paracellular shunt pathway and ionic conductance. *J. memb. Biol.*, **55**, 175–185.

Diamond, J. M. (1979) Osmotic water flow in leaky epithelia. *J. memb. Biol.*, **51**, 195–216.

Diamond, J. M. and Bossert, W. H. (1967) Standing-gradient flow. A mechanism for coupling of water and solute transport in epithelia. *J. gen. Physiol.*, **50**, 2061–2083.

Evans, D. H. (1979) "Fish" in *Comparative Physiology of Osmoregulation in Animals*, Vol. 1 (ed. Maloiy, G. M. O.), Academic Press, London, 305–390.

Fletcher, C. R. (1980) The relationship between active transport and the exchange diffusion effect. *J. theor. Biol.*, **82**, 643–661.

Forrest, J. N., MacKay, W. C., Gallagher, B. and Epstein, F. H. (1973) Plasma cortisol response to salt water adaptation in the American eel *Anguilla rostrata*. *Am. J. Physiol.*, **224**, 714–717.

Frizzell, R. A., Field, M. and Schultz, S. G. (1979) Sodium-coupled chloride transport by epithelial tissues. *Am. J. Physiol.*, **236**, F1–F8.

Hirano, T. (1980) "Effects of cortisol and prolactin on ion permeability of the eel oesophagus" in *Epithelial Transport in Lower Vertebrates* (ed. Lahlou, B.), Cambridge University Press, Cambridge, 143–149.

Hirano, T. and Mayer-Gostan, N. (1976) Eel oesophagus as an osmoregulatory organ. *Proc. Nat. Acad. Sci. U.S.A.*, **73**, 1348–1352.

Keys, A. B. and Willmer, E. N. (1932) "Chloride-secreting cells" in the gills of fishes with special reference to the common eel. *J. Physiol.*, **76**, 368–378.

Kirschner, L. B. (1977) "The sodium chloride excreting cells in marine vertebrates" in *Transport of Ions and Water in Animals* (ed. Gupta, B. L., Moreton, R. B., Oschman, J. L. and Wall, B. J.), Academic Press, London, 427–452.

Mayer, N., Maetz, J., Chan, D. K. O., Forster, M. E. and Chester Jones, I. (1967) Cortisol, a sodium excreting factor in the eel, *Anguilla anguilla* L., adapted to sea water. *Nature*, **214**, 1118–1120.

Maetz, J. (1971) Fish gills: Mechanisms of salt transfer in fresh water and in sea water. *Phil. Trans. roy. Soc. B.*, **262**, 209–249.

Maetz, J. (1974) "Adaptation to hypo- and hyperosmotic environments" in *Biochemical and Biophysical Perspectives in Marine Biology*, Vol. 1 (eds. Malins, D. C. and Sargent, J. R.), Academic Press, New York, 1–170.

Maetz, J. and Bornancin, M. (1975) Biochemical and biophysical aspects of salt excretion by chloride cells in teleosts. *Fortschritte der Zoologie*, **23**, 322–362.

Maetz, J. and Skadhauge, E. (1968) Drinking rates and gill ionic turnover in relation to external salinities in the eel. *Nature*, **217**, 371–373.

Motais, R. (1967) Les mécanismes d'échanges ioniques branchiaux chez les teléostéens. *Ann. Inst. oceanogr. Monaco*, **45**, 1–83.

Potts, W. T. W. (1976) "Ion transport and osmoregulation in marine fish" in *Perspectives in Experimental Biology*, Vol. 1, Zoology (ed. Spencer Davies, P.), Pergamon, Oxford, 65–75.

Potts, W. T. W. (1977) "Fish gills" in *Transport of Ions and Water in Animals* (eds. Gupta, B. L., Moreton, R. B., Oschman, J. L. and Wall, B. J.), Academic Press, London, 453–480.

Sargent, J. R., Bell, M. V. and Kelly, K. F. (1980) "The nature and properties of sodium ion plus potassium ion-activated adenosine triphosphatase and its role in marine salt secreting epithelia" in *Epithelial Transport in the Lower Vertebrates* (ed. Lahlou, B.), Cambridge University Press, Cambridge, 251–267.

Silva, P., Solomon, R., Spokes, K. and Epstein, F. H. (1977) Ouabain inhibition of gill Na-K-ATPase: Relationship to active chloride transport. *J. exp. Zool.*, **199**, 419–426.

Skadhauge, E. (1969) The mechanism of salt and water absorption in the intestine of the eel (*Anguilla anguilla*) adapted to waters of various salinities. *J. Physiol.*, **204**, 135–158.

Skadhauge, E. (1974) Coupling of transmural flows of NaCl and water in the intestine of the eel (*Anguilla anguilla*). *J. exp. Biol.*, **60**, 535–546.

Spring, K. R. and Hope, A. (1978) Size and shape of the lateral intercellular spaces in a living epithelium. *Science*, **200**, 54–58.

Spring, K. R. and Hope, A. (1979) Fluid transport and the dimensions of cells and interspaces of living *Necturus* gallbladder. *J. gen. Physiol.*, **73**, 287–305.

Stoff, J. S., Rosa, R., Hallac, R., Silva, P. and Epstein, F. H. (1979) Hormonal regulation of active chloride transport in the dogfish rectal gland. *Am. J. Physiol.*, **237**, F138–F144.

Chapter 6

Alvarado, R. H. (1979) "Amphibians" in *Comparative Physiology of Osmoregulation in Animals*, Vol. 1 (ed. Maloiy, G. M. O.), Academic Press, London, 261–303.

Babiker, M. M. and el Hakeem, O. H. (1979) Changes in blood characteristics and constituents associated with aestivation in the African lungfish *Protopterus annectens* Owen. *Zool. Anz.*, **202**, 9–16.

Delaney, R. G., Lahiri, S., Hamilton, R. and Fishman, A. P. (1977) Acid-base balance and plasma composition in the aestivating lungfish (*Protopterus*). Am. J. Physiol., **232**, R10–R17.

Deyrup, I. J. (1964) "Water balance and kidney" in *Physiology of the Amphibia* (ed. Moore, J. A.), Academic Press, New York, 251–328.

Edney, E. B. (1957) *The Water Relations of Terrestrial Arthropods*, Cambridge University Press, Cambridge.

Kramer, D. L., Lindsey, C. C., Moodie, G. E. E. and Stevens, E. D. (1978) The fishes and the aquatic environment of the central Amazon basin, with particular reference to respiratory patterns. *Can. J. Zool.*, **56**, 717–729.

Munro, A. F. (1953) The ammonia and urea excretion of different species of Amphibia during their development and metamorphosis. *Biochem. J.*, **54**, 29–36.

Chapter 7

Arlian, L. G. and Veselica, M. M. (1979) Water balance in insects and mites. *Comp. Biochem. Physiol.*, **64A**, 191–200.

Dunson, W. A. (1979) "Control mechanisms in reptiles" in *Mechanisms of Osmoregulation in Animals. Maintenance of Cell Volume* (ed. Gilles, R.), John Wiley & Sons, Chichester, 273–322.

Edney, E. B. (1977) "Transpiration in land arthropods" in *Transport of Ions and Water in Animals* (eds. Gupta, B. L., Moreton, R. B., Oschman, J. L. and Wall, B. J.), Academic Press, London, 657–672.

Fitzsimons, J. T. (1979) *The Physiology of Thirst and Sodium Appetite*, Cambridge University Press, Cambridge.

Kanwisher, J. W. (1966) Tracheal gas dynamics in pupae of the Cecropia silkworm. *Biol. Bull., Woods Hole mar. biol. Lab.*, **130**, 96–105.

Lasiewski, R. C. (1964) Body temperature, heart and breathing rate, and evaporative water loss in hummingbirds. *Physiol. Zoöl.*, **37**, 212–223.

MacFarlane, W. V., Morris, R. J. H. and Howard, B. (1963) Turn-over and distribution of water in desert camels, sheep, cattle and kangaroos. *Nature*, **197**, 270–271.

Maddrell, S. H. P. (1977) "Insect malpighian tubules" in *Transport of Ions and Water in Animals* (eds. Gupta, B. L., Moreton, R. B., Oschman, J. L. and Wall, B. J.), Academic Press, London, 541–569.

Maloiy, G. M. O. (1972) Renal salt and water excretion in the camel (*Camelus dromedarius*). *Symp. Zool. Soc. Lond.*, **31**, 243–259.

Maloiy, G. M. O., MacFarlane, W. V. and Shkolnik, A. (1979) "Mammalian herbivores" in *Comparative Physiology of Osmoregulation in Animals*, Vol. 2 (ed. Maloiy, G. M. O.), Academic Press, London, 185–209.

Mantel, L. H. (1979) "Terrestrial invertebrates other than insects" in *Comparative Physiology of Osmoregulation in Animals*, Vol. 1 (ed. Maloiy, G. M. O.), Academic Press, London, 175–218.

Minnich, J. E. (1979) "Reptiles" in *Comparative Physiology of Osmoregulation in Animals*, Vol. 1 (ed. Maloiy, G. M. O.), Academic Press, London, 391–641.

Minnich, J. E. and Piehl, P. A. (1972) Spherical precipitates in the urine of reptiles. *Comp. Biochem. Physiol.*, **41A**, 551–554.

Noble-Nesbitt, J. (1977) "Active transport of water vapour" in *Transport of Ions and Water in Animals* (eds. Gupta, B. L., Moreton, R. B., Oschman, J. L. and Wall, B. J.), Academic Press, London, 571–597.

Schmidt-Nielsen, K. (1964) *Desert Animals*, Oxford University Press, Oxford.

Schmidt-Nielsen, K., Schroter, R. C. and Shkolnik, A. (1980) Desaturation of the exhaled air in the camel. *J. Physiol.*, **305**, 74P–75P.

Shoemaker, V. H. and McClanahan, L. L. (1975) Evaporative water loss, nitrogen excretion and osmoregulation in Phyllomedusine frogs. *J. comp. Physiol.*, **100**, 331–345.

Shoemaker, V. H. and Nagy, K. A. (1977) Osmoregulation in amphibians and reptiles. *Ann. Rev. Physiol.*, **39**, 449–471.

Wall, B. J. (1977) "Fluid transport in the cockroach rectum" in *Transport of Ions and Water in Animals* (eds. Gupta, B. L., Moreton, R. B., Oschman, J. L. and Wall, B. J.), Academic Press, London, 599–612.

Wall, B. J. and Oschman, J. L. (1975) "Structure and function of the rectum in insects" in *Excretion* (ed. Wessing, A. R. E.), *Fortschr. Zool.*, **23**, 193–222.

Wall, B. J. and Oschman, J. L. (1979) "Insects" in *Comparative Physiology of Osmoregulation in Animals*, Vol. 1 (ed. Maloiy, G. M. O.), Academic Press, London, 221–260.

Chapter 8

Blair-West, J. R., Coghlan, J. P., Denton, D. A., Nelson, J. F., Orchard, E., Scoggins, B. A., Wright, R. D., Meyers, K. and Junqueira, L. C. U. (1968) Physiological, morphological and behavioural adaptation to a sodium deficient diet by wild native Australian and introduced species of animals. *Nature*, **217**, 922–928.

Blaustein, M. P. (1977) Sodium ions, calcium ions, blood pressure regulation and hypertension: a reassessment and a hypothesis. *Am. J. Physiol.*, **232**, C165–C173.

Dantzler, W. H. and Braun, E. J. (1980) Comparative nephron function in reptiles, birds and mammals. *Am. J. Physiol.*, **239**, R197–R213.

Dunson, W. A. (1976) "Salt glands in reptiles" in *Biology of the Reptilia*, Vol. 5 (eds. Gans, C. and Dawson, W. R.), Academic Press, 413–445.

Dunson, W. A., Packer, R. K. and Dunson, M. K. (1971) Sea snakes: an unusual salt gland under the tongue. *Science*, **173**, 437–441.

Fänge, R., Schmidt-Nielsen, K. and Osaki, H. (1958) The salt gland of the herring gull. *Biol. Bull. Woods Hole mar. biol. Lab.*, **115**, 162–171.

Hollenberg, N. K. (1980) Set point for sodium homeostasis: Surfeit, deficit and their implications. *Kidney Int.*, **17**, 423–429.

Kaul, R. and Hammel, H. T. (1979) Dehydration elevates osmotic threshold for salt gland secretion in the duck. *Am. J. Physiol.*, **237**, R355–R359.

Malvin, R. L. (1979) "Carnivores" in *Comparative Physiology of Osmoregulation in Animals*, Vol. 2 (ed. Maloiy, G. M. O.), Academic Press, London, 143–184.

Peaker, M. (1979) "Control mechanisms in birds" in *Mechanisms of Osmoregulation in Animals. Maintenance of Cell Volume* (ed. Gilles, R.), John Wiley & Sons, Chichester, 323–348.

Peaker, M. and Linsell, J. L. (1975) *Salt Glands in Birds and Reptiles*, Cambridge University Press, Cambridge.

de Wardener, H. E. and MacGregor, G. A. (1980) Dahl's hypothesis that a saluretic substance may be responsible for a sustained rise in arterial pressure: its possible role in essential hypertension. *Kidney Int.*, **18**, 1–9.

Willoughby, E. J. and Peaker, M. (1979) "Birds" in *Comparative Physiology of Osmoregulation in Animals*, Vol. 2 (ed. Maloiy, G. M. O.), Academic Press, London, 1–55.

Chapter 9

Abramow, M. (1979) "Control mechanisms in mammals" in *Mechanisms of Osmoregulation in Animals. Maintenance of Cell Volume* (ed. Gilles, R.), John Wiley & Sons, 349–412.

Andreoli, T. E. and Schafer, J. A. (1979) Effective luminal hypotonicity; the driving force for isotonic proximal tubular fluid absorption. *Am. J. Physiol.*, **236**, F89–F96.

Andreoli, T. E., Berliner, R. W., Kokko, J. P. and Marsh, D. J. (1978) Questions and replies: Renal mechanisms for urinary concentrating and diluting processes. *Am. J. Physiol.*, **235**, F1–F11.

Arendshorst, W. J. and Gottschalk, C. W. (1980) Glomerular ultrafiltration dynamics: euvolemic and plasma volume-expanded rats. *Am. J. Physiol.*, **239**, F171–F186.

Barajas, L. (1979) Anatomy of the juxtaglomerular apparatus. *Am. J. Physiol.*, **237**, F333–F343.

Blair-West, J. R., Coghlan, J. P., Denton, D. A., Nelson, J. F., Orchard, E., Scoggins, B. A., Wright, R. D., Meyers, K. and Junqueira, L. C. U. (1968) Physiological, morphological and behavioural adaptation to a sodium deficient diet by wild native Australian and introduced species of animals. *Nature*, **217**, 922–928.

Blair-West, J. R., Coghlan, J. P., Denton, D. A., Hardy, K. J., Scoggins, B. A. and Wright, R. D. (1979) Effect of adrenal arterial infusion of P-113 on aldosterone secretion in Na-deficient sheep. *Am. J. Physiol.*, **236**, F333–F341.

Brenner, B. M., Deen, W. M. and Robertson, C. R. (1974) "The physiological basis of glomerular ultrafiltration" in *Kidney and Urinary Tract Physiology* (ed. Thurau, K.), Butterworths, London & University Park Press, Baltimore, 335–356.

Brenner, B. M. and Rector, F. C. (1976) *The Kidney*, Vol. 1, W. B. Saunders Company, Philadelphia.

Brenner, B. M. and Stein, J. H. (eds.) (1978) *Sodium and Water Homeostasis*, Churchill Livingstone, New York.

Brenner, B. M. and Stein, J. H., eds. (1980) *Hormonal Function and the Kidney*, Churchill Livingstone, New York.

Churchill, P. C., Churchill, M. C. and McDonald, F. (1978) Renin secretion and distal tubule sodium in rats. *Am. J. Physiol.*, **235**, F611–F616.

Davis, J. O. and Freeman, R. H. (1976) Mechanisms regulating renin release. *Physiol. Rev.*, **56**, 1–56.

Gunther, R. A. and Rabinowitz, L. (1980) Urea and renal concentrating ability in the rabbit. *Kidney Int.*, **17**, 205–222.

Harvey, R. J. (1974) *The Kidneys and the Internal Environment*, Chapman & Hall, London.

Jamison, R. L., Bennett, C. M. and Berliner, R. W. (1967) Countercurrent multiplication by the thin loop of Henle. *Am. J. Physiol.*, **212**, 357–366.

Jamison, R. L., Roinel, N. and de Rouffignac, C. (1979) Urinary concentrating mechanism in the desert rodent, *Psammomys obesus. Am. J. Physiol.*, **236**, F448–F453.

Kokko, J. P. and Rector, F. C. (1972) Countercurrent multiplication system without transport in inner medulla. *Kidney Int.*, **2**, 214–223.

McFarland, W. N. and Wimsatt, W. A. (1969) Renal function and its relation to the ecology of the Vampire Bat, *Desmodus rotundus. Comp. Biochem. Physiol.*, **28**, 985–1006.

Moffatt, D. B. (1975) *The Mammalian Kidney*, Cambridge University Press, Cambridge.

Moore, L. C. and Marsh, D. J. (1980) How descending limb of Henle's loop permeability affects hypertonic urine formation. *Am. J. Physiol.*, **239**, F57–F71.

Morris, D. J., Berck, J. S. and Davis, R. P. (1973) The physiological response of aldosterone in adrenalectomised and intact rats and its sex dependence. *Endocrinology*, **92**, 989–993.

Peach, M. J. (1977) Renin-angiotensin system: Biochemistry and mechanisms of action. *Physiol. Rev.*, **57**, 313–370.

Pitts, R. F. (1968) *Physiology of the Kidney and Body Fluids*, 2nd Edn., Year Book Medical Publishers, Chicago.

Robertson, G. L., Athar, S. and Shelton, R. L. (1977) "Osmotic control of vasopressin function" in *Disturbances in Body Fluid Osmolality* (eds. Andreoli, T. A., Grantham, J. J. and Rector, F. C.), American Physiological Society, Bethesda, 125–148.

Verney, E. B. (1947) The antidiuretic hormone and the factors which determine its release. *Proc. roy. Soc. B*, **135**, 25–106.

de Wardener, H. E. (1978) The control of sodium excretion. *Am. J. Physiol.*, **235**, F163–F173.

Wright, F. S. and Briggs, J. P. (1979) Feedback control of glomerular blood flow, pressure and filtration rate. *Physiol. Rev.*, **59**, 958–1006.

Chapter 10

Crisp, D. J., Davenport, J. and Gabbott, P. A. (1977) Freezing tolerance in *Balanus balanoides. Comp. Biochem. Physiol.*, **57A**, 359–361.

Davenport, J. (1979) Cold resistance in *Gammarus duebeni* Liljeborg. *Astarte*, **12**, 21–26.

Denton, E. J., Gilpin-Brown, J. B. and Shaw, T. I. (1969) A buoyancy mechanism found in cranchid squid. *Proc. Roy Soc. Lond. B*, **174**, 261–269.

Denton, E. J. and Gilpin-Brown, J. B. (1973) Flotation mechanisms in modern and fossil cephalopods. *Adv. mar. Biol.*, **11**, 197–268.

Maetz, J. and de Renzis, G. (1978) "Aspects of the adaptation of fish to high external alkalinity: comparison of *Tilapia grahami* and *T. mossambica*" in *Comparative Physiology: Water, Ions and Fluid Mechanics* (eds. Schmidt-Nielsen, K., Bolis, L. and Maddrell, S. H. P.), Cambridge University Press, Cambridge, 213–228.

McWilliams, P. G. and Potts, W. T. W. (1978) The effects of pH and calcium concentrations on gill potentials in the brown trout, *Salmo trutta. J. comp. Physiol.*, **26**, 277–286.

Oikari, A. (1975) Hydromineral balance in some brackish-water teleosts after thermal acclimation, particularly at temperatures near zero. *Ann. Zool. Fennici*, **12**, 215–229.

Packer, R. K. and Dunson, W. A. (1970) Effects of low environmental pH on blood pH and sodium balance of brook trout. *J. exp. Zool.*, **174**, 65–72.

Phillips, J. E. and Bradley, T. J. (1977) "Osmotic and ionic regulation in saline-water mosquito larvae" in *Transport of Ions and Water in Animals* (eds. Gupta, B. L., Moreton, R. B., Oschman, J. L. and Wall, B. J.), Academic Press, London.

Scholander, P. F., Dam, L. Van, Kanwisher, J. W., Hammel, H. J. and Gordon, M. S. (1957) Supercooling and osmoregulation in Arctic fish. *J. Cell comp. Physiol.*, **49**, 5–24.

Skadhauge, E., Lechène, C. P. and Maloiy, G. M. O. (1980) "*Tilapia grahami*: role of intestine in osmoregulation under conditions of extreme alkalinity" in *Epithelial Transport in the Lower Vertebrates* (ed. Lahlou, B.), Cambridge University Press, Cambridge, 133–142.

Vernberg, F. J. and Silverthorn, S. U. (1979) "Temperature and osmoregulation in aquatic species" in *Mechanisms of Osmoregulation in Animals, Maintenance of Cell Volume* (ed. Gilles, R.), John Wiley & Sons, Chichester, 273–377.

Vries, A. L. de and Wohlschlag, D. E. (1969) Freezing resistance in some Antarctic fishes. *Science*, **163**, 1073–1075.

Williams, R. J. (1970) Freezing tolerance in *Mytilus edulis. Comp. Biochem. Physiol.*, **35**, 145–161.

Chapter 11

Bettison, J. C. and Davenport, J. (1976) Salinity preference in gammarid amphipods with special reference to *Marinogammarus marinus* (Leach). *J. mar. Biol. Ass. U.K.*, **56**, 135–142.

Dantzler, W. H. (1977) "*In vitro* microperfusion" in *Transport of Ions and Water in Animals* (eds. Gupta, B. L., Moreton, J. C., Oschman, J. L. and Wall, B. J.), Academic Press, London, 57–82.

Davenport, J. (1976) A comparative study of the behaviour of some balanomorph barnacles exposed to fluctuating sea water concentrations. *J. mar. Biol. Ass. U.K.*, **56**, 889–907.

Erasmus, D. A. (ed.) (1978) *Electron Probe Microanalysis in Biology*, Chapman & Hall, London.

Giebisch, G. (1977) "Micropuncture techniques" in *Transport of Ions and Water in Animals* (eds. Gupta, B. L., Moreton, R. B., Oschman, J. L. and Wall, B. J.), Academic Press, London, 29–55.

Gupta, B. L., Hall, T. A. and Moreton, R. B. (1977) "Electron probe X-ray analysis" in *Transport of Ions and Water in Animals* (eds. Gupta, B. L., Moreton, R. B., Oschman, J. L. and Wall, B. J.), Academic Press, London, 83–143.

Lechène, C. P. and Warner, R., eds. (1979) *Microbeam Analysis in Biology*, Academic Press, New York.

Little, C. (1977) "Microsample analysis" in *Transport of Ions and Water in Animals* (eds. Gupta, B. L., Moreton, R. B., Oschman, J. L. and Wall, B. J.), Academic Press, London, 15–28.

McLusky, D. A. (1970) Salinity preference in *Corophium volutator. J. mar. Biol. Ass. U.K.*, **50**, 747–752.

Motais, R. (1967) Les mécanismes d'échanges ioniques branchiaux chez les teleostéens. *Ann. Inst. oceanogr. Monaco*, **45**, 1–83.

Ramsey, J. A. and Brown, R. H. J. (1955) Simplified apparatus and procedure for freezing-point determination upon small volumes of fluid. *J. scient. Instr.*, **32**, 372–375.

Trueman, E. R. (1967) Activity and heart rate of bivalve molluscs in their natural habitat. *Nature*, **214**, 832–833.

Index